SCIENCE IN TRADITIONAL CHINA

Science in Traditional China
A Comparative Perspective

Joseph Needham

Harvard University Press
CAMBRIDGE, MASSACHUSETTS

The Chinese University Press
HONG KONG

Library of Congress Cataloging in Publication Data

Needham, Joseph, 1900-
  Science in traditional China.

  1. Science—China—History. I. Title.
Q127.C5N46   1981   509.51   81-6962
ISBN 0-674-79438-9 Harvard University Press (cloth)
ISBN 0-674-79439-7 (paper)   AACR2

Contents

Illustrations

Preface

The lectures which are grouped together in this book were given originally as the Second Series of Ch'ien Mu Lectures[a] at New Asia College, The Chinese University (Chung Wên Ta Hsüeh) at Shatin, Hong Kong. I have the warmest memories of everything about that visit, the kindness and welcome of all my academic colleagues, the charm and enthusiasm of the student audiences, the unusual beauty of the Shatin campus and its surroundings, and the ever-present thrill of proximity to a great Chinese city. It is certainly my hope that many of the facts disclosed in these lectures will serve to bring readers both Eastern and Western to a more just appreciation of the role of science, technology, and medicine in the Chinese culture-area throughout history.

It is now forty-three years since I was first stimulated to take up the study of Chinese language and culture, not knowing at that time whether I would ever put it to great use. Now many volumes of the *Science and Civilisation in China* series have been published, but many more remain to be finalised and passed through press. We divide the volumes into "heavenly" and "earthly." The former are those of the original project, as it was planned when we blithely went through the spectrum of the sciences and drew up a general plan. What we could not tell at that time was the amount of space which would have to be devoted to the different forms of science, pure and applied; and it was this which made it necessary for some of the "heavenly" volumes to come out in several parts. These indeed are the "earthly" physical volumes. About eleven of these are now either published or going through the press, which leaves about eight or nine yet to be done. Since I am now in my eighty-first year, we generally say that if I can go on

[a] 錢賓四先生學術文化講座

working till ninety, I have a sporting chance of seeing the voyage pretty nearly into port. I am glad to say that many of the volumes of the future are already in advanced draft form, though there is much editing and polishing yet to be done. Moreover, we now have some twenty collaborators scattered throughout the world, who together are undertaking far more than any one or two individuals could have accomplished.

This leads me to say that nothing would ever have been possible without the equal collaboration of Chinese friends. My view is that no group of Chinese, or of Westerners, could do it alone—the range of knowledge and expertise is far too great. So first I would like to commemorate Lu Gwei-Djen, my oldest Chinese friend, now Associate Director of the East Asian History of Science Library in Cambridge; and secondly Wang Ling (Wang Ching-Ning), my first collaborator, who worked with me for nine consecutive years in rather cramped quarters in Caius College. But many others should be named, for example Ho Ping-Yü, successively of Singapore, Kuala Lumpur, Brisbane, and Hong Kong, Lo Jung-Pang of California, Huang Jen-Yü of New York, Ch'ien Ts'un-Hsün of Chicago, and most recently Ch'ü Chih-Jen, working on the ceramics section. I cannot name everyone, but they have not all been Chinese. Among Europeans I would like to mention Kenneth Robinson of Oxford and Sarawak, Janusz Chmielewski of Poland, and Georges Metailié of France. Also from across the Atlantic came Nathan Sivin, now of Philadelphia, Robin Yates of Harvard, and Ursula Franklin of Toronto. Thus, as it has turned out, we have been a remarkably international group, a fact in itself of good augury for the future, since whatever else our work is, it surely has to be considered an essay towards the better mutual understanding of the peoples, and so a step along the road to general peace and friendship.

When I look back forty years, and again to the time when I was working in Unesco, and used to read the *Tso Chuan* while in my bath of an evening, I feel very impressed with the fact that in those days we had nothing but the classics to go on. By this I have particularly in mind the works of the great sinologists of the last century and the first half of the present one—men like Chavannes, Couvreur, Pelliot, Friedrich Hirth, Otto Franke, H. A. Giles, and so on. The number of scholarly translations was then very few as compared to today. We collected all these books for our library, but at the present time, what a difference; our new bookshelf continually groans with treatises and monographs of all kinds—hydraulic engineering in the Sung, naval architecture from Han to Ming, medical ethics in classical China, etc., etc. The list is endless. I think possibly we may ourselves take some credit in having inspired this flow of valuable works, unless indeed we were just part of a historical movement which impelled Westerners to a more thorough study of Chinese culture. But in China itself since the revolution, there has also been a great flowering of scholarship. Archaeologists in the West complain of being snowed under by the output of the Chinese archaeological reports. New books have appeared on all aspects of the history of science, technology, and medicine, and really wonderful discoveries have been made. I take much pleasure in looking back to think that we were part of this great wave, perhaps a pioneering part.

Quite recently, the editors of that excellent English historical journal *Past and Present* printed a symposium upon our work. They said they had had difficulty in finding contributors because almost nobody in the Western world who had qualifications both in Chinese and in the history of science had failed to be incorporated into our group; but in fact they found excellent writers— Mark Elvin, Willard Peterson, Ulrich Libbrecht, and Christopher Cullen. It did, as the chairman said, produce an almost perfect blend of sympathy and critical appraisal. But I was rather amused by some bitter-sweet remarks. For example, a comparison was made with Toynbee and Frazer, suggesting a hint or implication that a certain element of the subjective had crept into our presentations. This however I would willingly accept, because I think that everyone who undertakes a big inter-cultural job like this must naturally project his own system of beliefs in doing so—it is his opportunity to preach (and I use the word quite advisedly) to his own and later generations. If sometimes we have written like barristers pleading a case, or sometimes over-emphasise the Chinese contributions, it has been consciously to redress a balance which in the past tilted over much too far on the other side. We were out to redress a secular injustice and misunderstanding.

In the preface to one of the volumes of *Science and Civilisation in China*, we wrote the following words, which I still find entertaining: "Some time ago," we said, "a not wholly unfriendly critic of our previous volumes wrote, in effect: This book is fundamentally unsound for the following reasons. The author's belief is (1) that human social evolution has brought about a gradual increase in man's knowledge of Nature and control of the external world, (2) that this science is an ultimate value, and with its applications forms today a unity into which the comparable contributions of different civilisations ... all have flowed and flow as rivers to the sea, (3) that along with this progressive process human society is moving towards forms of ever greater unity, complexity, and organisation. We recognise these invalidating theses as indeed our own, and if we had a door like that of Wittenberg long ago, we would not hesitate to nail them to it.'' I can now say that our critic was the late Arthur Wright, actually a great friend, but one whose Buddhist unworldliness and political pessimism had led him to a world-view very different from our own.

All in all, our volumes have been essentially reconnaissances of the most exciting character. We never intended them as ''the last word'' on any subject, for such definitiveness would have been quite impossible in the state of the art, and indeed is still impossible today. But the excitement of the search remains—the recognition of ideas and thoughts, the discovery of unexpected priorities under unfamiliar terminology, the welcome of unexpected forerunners and the admiration for their work, the understanding of inventions and techniques which had not been unearthed before. It has all been an exciting reconnaissance. One would like to say, paraphrasing the *Tao Tê Ching*, that ''when the Great Way declined,'' then adjectives like ''competence'' and ''incompetence'' began to fill the air. Then distinctions between ''sound'' and ''unsound'' began to appear. Let the ultimate balance be cast up by the centuries to come. What we know is that we have met with our Chinese brothers and sisters in the fields of science, technology, and medicine during the past twenty-five centuries, and though we can never speak with them, we can often read their words, and we have sought to give them their meed of honour.

JOSEPH NEEDHAM
21 January 1981

1
Introduction

I have no objection this evening to telling you about the rather curious sequence of historical circumstances which led eventually to the volumes in the *Science and Civilisation in China* series (SCC). For this purpose we have to go back to the end of the First World War when I arrived in Cambridge at Caius College, that home of learning of which forty-seven years later I. was to become Master. Since my father was a physician and an early specialist in anaesthesia, I was destined for medicine, but in the very first few years I was spiritually shunted away from this by the charismatic lectures of "Hoppy," in other words, Sir Frederick Gowland Hopkins,[1] O.M., P.R.S.—like iron filings responding to the magnet, we all became biochemists under him. "Hoppy" was *in loco parentis* to us, and the only other person I feel that about was Charles Singer,[2] perhaps the greatest British historian of science in the first half of this century.

So under Hoppy I became a biochemist, and became interested in organic syntheses, and then I found that the developing hen's egg was nothing less than a marvellous chemical factory in which a great deal is synthesized during its three weeks of activity. However, it was one thing to trace the formation of substances like inositol or ascorbic acid during embryonic development, but quite another to face up to the problems of the morphological construction of the embryo from the original fertilised egg cell. That was how I became deeply involved in agitating philosophical questions. In the very same year that *Chemical Embryology* was published, the first experiments were made which showed that the primary induction centre in the amphibian embryo is stable to

[1] Sir Frederick Gowland Hopkins, 1861–1947, "the father of British biochemistry," Professor at Cambridge.

[2] Charles Singer, eminent English historian of science and medicine.

boiling. This study of what I called "morphogenetic hormones" led to another book, *Biochemistry and Morphogenesis*, ten years later.

So I was, in a way, myself part of a historical story, almost a character in a historical play; but it so happened that I had had a passion for history ever since my schooldays, and experimental science alone was never able to satisfy me, and thus it came about that I felt the necessity of preluding *Chemical Embryology* with a long account of the history of embryology from the beginnings, so far as I could describe them, down to about 1800. And that was where Charles Singer came in again, because I suppose I can say he was the only real teacher in the history of science that I ever had, though in fact I never heard a single formal lecture by him. But I was a personal friend and received all sorts of good advice, with many a clue as to where to look for things. Over many years I was accustomed to go and stay in his house on the Cornish coast, stacked as it was from floor to ceiling with a marvellous library. Now, in the history of embryology which I wrote in those days I was, of course, particularly interested in the pioneers of chemical embryology like Walter Needham,[3] + 1668, one of my own family and a foundation fellow of the Royal Society, and also Sir Thomas Browne,[4] in his 17th-century laboratory in Norwich, trying to probe the secrets of egg white and egg yolk with the chemical methods of his time. So history was battling with science as the form of experience which would claim most of my time, until in 1937 a new induction phenomenon occurred. Art and religion, I ought to add, were always in the picture, too, but not so dominatingly.

The induction phenomenon to which I refer was the appearance in Cambridge in that year of several young Chinese research workers who came to take their doctorates. The most compelling influence among these friends was incarnated, I may say, in Lu Gwei-Djen,[a] who now, forty-two years afterward, is my chief collaborator and the associate director of our library. These friends exercised two primary effects on me: first they inspired me to learn

[3] Walter Needham, +17th-century doctor and embryologist, one of the founding Fellows of the Royal Society.

[4] Sir Thomas Browne, + 1605—1682, famous English doctor and author whose works include *Religio Medici*.

[a] 魯桂珍

their language, and second they raised the question of why modern science originated only in Europe.

On the matter of language, it is a well-known fact that occasionally Westerners are struck down by a blinding light, like Saint Paul on the road to Damascus, with the feeling that they must learn this language with its marvellous script or else burst. That was perhaps not so surprising, but the effect in the mental world was a very striking one, because I found that the more I got to know these friends from China the more exactly like my own their minds seemed to be, certainly in their intellectual capacity; and this raised in an acute form, therefore, the question of why modern science, the "new or experimental" philosophy of the time of Galileo, had arisen only in European culture and not in Chinese or Indian.

Many years later, when I had learned a lot more about these things, I realised that there was a second question hiding behind that first one: namely, how could it be that the Chinese civilisation had been much more effective than the European in finding out about Nature and using natural knowledge for the benefit of mankind for fourteen centuries or so before the scientific revolution?

But still there might not have been a "take" (like a failed vaccination) if it had not fallen to my lot to be asked to become Scientific Counsellor at the British Embassy in Chungking throughout the Second World War. Four years in China sealed my fate. After that it was impossible to think of doing anything else but a book on the history of science, technology, and medicine in China, something which had not previously existed in any Western language. I say "a book," and we did at first think of it in terms of a single slim volume, but the unrolling of history decreed that it was not to be like that. Starting out blithely, we ran through the spectrum of the sciences dividing up the universe into seven volumes, and this arrangement we still keep to; but the exigencies of the work and the sheer mass of the material have meant that the different volumes in fact each have to come out in as many as half a dozen parts, so that in the end the final work will probably run to about twenty volumes.

When I started out on this work my sinological friends in Cambridge did not believe that I would find anything interesting at all; they even doubted whether Chinese culture had ever had any science, technology, or medicine significant for the world. Gustave Haloun,[5] Professor of Chinese at the time in Cambridge, following that great sinologist, Friedrich Hirth,[6] used to talk rather wistfully about the *realia*, the actual things, which one ought to know about in order to understand the texts—the ploughs, the pottery and porcelain, the papermakers' tools, and so on—but that was as far as it went.

Only after I got to China did I find everywhere scientists, doctors, and engineers who themselves took a great interest in the history of their own subjects in their own ideographic culture, and they proved ready and willing to indicate to me what were the most important Chinese books which should be purchased and studied. The result was the opening up of a veritable gold mine, a cornucopia which would have surprised all the older sinologists, as indeed it surprised me, and perhaps also the classical Chinese scholars too.

By the time I got back to Cambridge after the war I had acquired my first chief collaborator, Wang Ching-Ning,[a7] from the History Institute of Academia Sinica, and when, rather more than twenty years later, he left to take up a research post in Australia, I was able to persuade my oldest friend, Lu Gwei-Djen, to rejoin me from Unesco and press on with the work. If all of us could live to one hundred fifty years each we might have hoped to achieve the grand design single-handedly, but since that cannot be, we have acquired since then many collaborators, and we now have twenty or more scattered over the world. I mention only a very few: Lo Jung-Pang[b8] in California, Ch'ien Ts'un-Hsün[c9] in Chicago, Ursula Franklin[10]

[5] Gustave Haloun, Professor of Chinese at Cambridge between 1938 and his death in 1951.

[6] Friedrich Hirth wrote many articles on Chinese economic and cultural history. A German, he was active at about 1900.

[7] Wang Ching-Ning, Professor of Chinese at the Australian National University, collaborator in many sections of SCC.

[8] Lo Jung-Pang, Emeritus Professor of History at the University of California, collaborator in SCC for the sections on military technology, the salt industry, and deep drilling.

[9] Ch'ien Ts'un-Hsün, Emeritus Professor at the University of Chicago; collaborator in SCC for the history of paper and printing.

[10] Ursula Franklin, Professor of Metallurgy at the University of Toronto, collaborator in SCC for the history of nonferrous metallurgy.

[a] 王靜寧   [b] 羅榮邦   [c] 錢存訓

in Toronto, Huang Jen-Yü[a][11] in New York, Ho Ping-Yü[b][12] in Brisbane, Ch'ü Chih-Jen[c][13] not far from us here, and so on. I may well not live to see the proofs of the last volume myself, but the future of the project is assured, and beyond that we anticipate that our East Asian History of Science Library, as it now is, will be established in the new building it so much requires in the compound of the new foundation of Robinson College in Cambridge, of which Lu Gwei-Djen is a Fellow and I am a Trustee.

Well, we have been to a certain extent pioneers. Of course, in the fairly recent past there have been great historians of mathematics in China and Japan, such as Li Yen,[d][14] Ch'ien Pao-Tsung,[e][15] and Mikami Yoshio,[f][16] and great historians of astronomy, like Leopold de Saussure,[17] Ch'ên Tsun-Kuei,[g][18] and Chu K'o-Chen.[h][19] There are still living today profound exponents of Taoism, alchemy, and early chemistry, like Ch'ên Kuo-Fu[i][20] and Wang Ming.[j][21] Botany and agriculture have had their brilliant historians, such as Hsia Wei-Ying,[k][22] Shih Shêng-Han,[l][23] and

[11] Huang Jen-Yü, sometime Professor of East Asian History at the State University of New York, collaborator in SCC for economic and social history.

[12] Ho Ping-Yü, Professor of Chinese at Griffith University, Brisbane, collaborator in SCC for the history of alchemy, early chemistry, pharmacology, and gunpowder. [Now at Hong Kong University, Hong Kong. — Ed.]

[13] Ch'ü Chih-Jen, Reader in the History of Art at The Chinese University of Hong Kong, collaborator in SCC for ceramics technology.

[14] Li Yen, eminent historian of Chinese mathematics.

[15] Ch'ien Pao-Tsung, eminent historian of Chinese mathematics.

[16] Mikami Yoshio, eminent historian, wrote *The Development of Mathematics in China and Japan* (1913).

[17] Leopold de Saussure, French naval officer and sinologist whose works include *Les Origines de l'Astronomie Chinoise.*

[18] Ch'ên Tsun-Kuei, eminent historian of Chinese astronomy.

[19] Chu K'o-Chen, former Vice-President of Academia Sinica, wrote widely in astronomy, the history of climatology, and meteorology, and on the calendar and the inhibited development of modern science in China.

[20] Ch'ên Kuo-Fu, Professor of Chemistry at Tientsin University and authority on Taoist literature and history of alchemy.

[21] Wang Ming, authority on Taoism and its influence on scientific techniques.

[22] Hsia Wei-Ying, eminent historian of botany in China.

[23] Shih Shêng-Han, eminent historian of Chinese agriculture.

[a] 黃仁宇   [b] 何丙郁   [c] 屈志仁   [d] 李儼   [e] 錢寶琮

[f] 三上義夫   [g] 陳遵嬀   [h] 竺可禎   [i] 陳國符   [j] 王明

[k] 夏緯瑛   [l] 石聲漢

Amano Motonosake.[a][24] Medicine has had many eminent expositors; one thinks of Li T'ao[b][25] and Ch'ên Pang-Hsien.[c][26] Less has been done on the history of technology and engineering, though Hu Tao-Ching[d][27] has won great merit by his studies of Shen Kua's[e][28] *Mêng Ch'i Pi T'an*[f] and Berthold Laufer[29] took up many fascinating aspects of applied science.

But somehow or other no one before us felt the vocation, or perhaps I ought to call it the obsession, to gather together in encyclopaedic form all that was known of the history of *all* the sciences, together with technology and medicine, in Chinese culture through the ages, and compare it step by step with what was known and achieved in the other parts of the Old World, in the cultures of Europe, Islam, India, and Persia. Only in this way, to be sure, could any estimate be made of the indebtedness of the civilisations to one another and of the facilitations or inhibitions which affected intercourse between them. For example, we hope to show in volume V, part 4, and in a later lecture in this present series, that the conception of the life-elixir, originating in China and only in China, passed first to the Arabs, then to the Byzantines, and lastly to the Franks or Latins in the time of Roger Bacon,[30] setting on foot the whole movement of chemical medicine. When the great Paracelsus von Hohenheim,[31] before the end of the 15th century, affirmed that "the business of alchemy is not to make gold, but to prepare remedies for human ills," he was

[24] Amano Motonosake, eminent Japanese historian of hydraulic works, agricultural technology, the social aspects of agriculture, and specific agricultural texts.

[25] Li T'ao, eminent historian of medicine.

[26] Ch'ên Pang-Hsien, eminent historian of Chinese medicine.

[27] Hu Tao-Ching, eminent contemporary scholar in the history of science, edited the *Mêng Ch'i Pi T'an*.

[28] Shen Kua, scholar-official of scientific bent, born in + 1030, wrote *Mêng Ch'i Pi T'an* in 1086.

[29] Berthold Laufer, outstanding German sinologist who lived from 1874–1934, wrote many articles on Chinese cultural history.

[30] Roger Bacon, + 1214–1294, English philosopher of scientific interests.

[31] Paracelsus (von Hohenheim), + 1493–1541, Swiss physician and greatest of the iatro-chemists.

[a] 天野元之助   [b] 李濤   [c] 陳邦賢   [d] 胡道靜   [e] 沈括
[f] 夢溪筆談

standing in a direct line of descent from Li Shao-Chün[a][32] and Ko Hung[b][33]—for death itself, the supreme of human ills, was what they believed that man could be cured of.

All pioneers are liable to be rather isolated in their generation, and we have been no exception. Faculties of Oriental Studies have never wanted to have much to do with us, mainly I think because they are traditionally composed of humanists, philologists, and linguists. They have never had the time to get to know anything much about science, technology, and medicine, and they feel they can hardly begin now. By the same token, strangely enough, a similar wall of glass has separated to us from departments of the history of science, for, generally speaking, they are interested primarily in European post-Renaissance science, partly because they lack all access to primary sources in other languages. Sometimes they are interested in Greek science as well, very rarely in mediaeval or Arabic science. The last thing they want to hear about is non-European science, and that is partly because of their strongly Europocentric vision. The unspoken assumption is that because distinctively modern science originated only in Europe, the only interesting ancient and mediaeval sciences must also have been European. This glaring *non sequitur* still dominates the intellectual outlook in the West, in spite of all that has been done by enlightened comparative historians of technology, like Lynn White,[34] who have shown over and over again the indebtedness of traditional Europe to discoveries and inventions made in the more easterly parts of the Old World.

But we must not complain too much, for we have been honoured in our time by that fabulous body of orientalists, the Royal Asiatic Society in Calcutta, and the historians of science have crowned our work with the laurels associated with the names of Leonardo da Vinci and George Sarton[35] and with Dexter's Plaque.[36]

---

[32] Li Shao-Chün, Han alchemist, floruit −133.

[33] Ko Hung, Chin scholar and Taoist alchemist, ca. +280−350.

[34] Lynn White, eminent American historian of technology, who has written on the interactions of religion, social change, and technology in mediaeval times.

[35] George Sarton, 1884−1956, outstanding American historian of science.

[36] Dexter Plaque, award given for the history of chemistry.

[a] 李少君   [b] 葛洪

Now, why continuous progress has been so much more evident in some fields than others is one of the mysteries of historiography. Where the arts are concerned, there is indeed a certain incommensurability between the civilisations, and little continuous development can be found among them. I suppose there can hardly have been a better sculptor than Pheidias[37] in any age before or since. No poet, either before or after, has outdone Tu Fu[a] or Pai Chü-I,[b38] and few playwrights at any time have written decisively better than Shakespeare, but where science, technology, and medicine are concerned, there *is* a clear increase in man's knowledge and power through the centuries. Nature has remained approximately the same since man began, and we believe that the growth of man's knowledge about Nature has been one single epic rise from the beginning until now, and now is not the end. Unquestionably Chang Hêng[c39] knew more about seismology than Xenocrates;[40] in time measurement Su Sung[d41] outdid Vitruvius;[42] and although Isaac Newton penetrated far indeed into the world of Nature, Einstein penetrated further. Hence we cannot in the last resort accept Oswald Spengler's[43] idea of every civilisation being complete in itself, having no relation with any other, and living out its own life cycle independently, like a plant, an animal, or an individual human being. This may be true for artistic styles, but it is only partly true for religions and philosophies, and assuredly untrue for science, technology, and medicine. Here, we believe, humanity has always been marching in one column, and though the particular systems of natural philosophy may have been ethnically bound, as I will mention in a minute, and therefore untranslatable, the actual understanding and control over Nature has

[37] Pheidias (or Phidias), −5th-century Greek sculptor.
[38] Tu Fu (+712−770) and Pai Chü-I (+772−846), two of China's most celebrated poets.
[39] Chang Hêng, +78−139, Han astronomer-mathematician and inventor of the seismograph.
[40] Xenocrates, −395 to −314, Greek philosopher.
[41] Su Sung, +1020−1101, scholar-official and astronomer whose book on the water-powered drive mechanism for an armillary sphere and celestial globe (*Hsin I Hsiang Fa Yao*[c]) revealed six centuries of clockmaking in China before Europe.
[42] Vitruvius, −1st-century Roman engineer, wrote a treatise on architecture.
[43] Oswald Spengler, German historical philosopher and author of *The Decline of the West* (1918).

[a] 杜甫   [b] 白居易   [c] 張衡   [d] 蘇頌   [e] 新儀象法要

been passed down from mind to mind across all barriers to build up what the early Royal Society called the great edifice of "true natural knowledge." Thus in the *Naturwissenschaften* there has been progress, and the *Geisteswissenchaften* have to a large extent participated in it, too. For example, to prove that the Roman Decretals[44] were forgeries, to determine the true date of the *Hermetic Books*,[45] and to settle the sources and dates of the *Lieh Tzŭ*[a] were real and permanent advances in knowledge. So we say *Floreat Scientia*.

I think I had better now define modern and mediaeval science a little bit more clearly, because we must make an important distinction between the two. The sciences of the mediaeval world were in fact tied closely to their ethnic environment, and it was difficult if not impossible for people of those different environments to find any common basis of discourse. For example, if Chang Hêng had tried to talk to Vitruvius about the Yin[b] and the Yang[c] or the Five Elements, he would not have got very far, even if they could have understood each other at all. But that did not mean that it was impossible for inventions of great sociological importance to pass from one civilisation to another, and that they did, right through the Middle Ages.

When we say that modern science developed only in Western Europe in the time of Galileo during the Renaissance and during the scientific revolution, we mean, I think, that it was there alone that there developed the fundamental bases of *modern* science, such as the application of mathematical hypotheses to Nature, and the full understanding and use of the experimental method, the distinction between primary and secondary qualities, and the systematic accumulation of openly published scientific data. Indeed, it has been said that it was in the time of Galileo that the most effective method of discovery about Nature was itself, and I think that is still quite true.

Nevertheless, before the river of Chinese science flowed, like all other such rivers, into the sea of modern science, China had seen remarkable achievements in many directions. For example, take

---

[44] Roman Decretals, forged documents which purported to prove the descent of temporal power from the Roman emperors to the Pope or Latin Patriarch.

[45] *Hermetic Books* religious texts of ancient Alexandrian date.

[a] 列子   [b] 陰   [c] 陽

mathematics: decimal place value and a blank space for the zero had begun in the land of the Yellow River earlier than anywhere else, and a decimal metrology had gone along with it. By the −1st century, Chinese artisans were checking their work with sliding calipers decimally graduated. Chinese mathematical thought was always deeply algebraic, not geometrical,[46] and in the Sung[a] and Yüan[b] the Chinese led the world in the solution of equations, so that the triangle called by the name of Blaise Pascal[47] was already old in China in + 1300.[48] We often find examples of this sort. The system of linked and pivoted rings which we know as the Cardan suspension, after Jerome Cardan,[49] really ought to be called Ting Huan's[c50] suspension because it had been used in China a whole thousand years before the time of Cardan. As for astronomy, we need only say that the Chinese were the most persistent and accurate observers of celestial phenomena anywhere before the Renaissance. Although geometrical planetary theory did not develop among them, they conceived an enlightened cosmology, mapped the heavens using our modern coordinates (and not the Greek ones), and kept records of eclipses, comets, novae, meteors, sun-spots, and so on that are used by radio astronomers down to this very day. A brilliant development of astronomical instruments also occurred, including the invention of the equatorial mounting and the clock drive, and this development was in close dependence on the contemporary capacities of Chinese engineers. I have already mentioned seismology as a case in point because the world's first seismograph was built by Chang Hêng,[d] as we all know, probably in about + 130.

Three branches of physics were particularly well developed in ancient and mediaeval China: optics, acoustics, and magnetism. This was in striking contrast with the West, where mechanics and

---

[46] For details, see SCC, vol. III (Cambridge University Press, 1959), section 19, especially pp. 112−146; also ibid., pp. 23−24.

[47] Blaise Pascal, + 1623−1662, French mathematician, physicist, and moralist.

[48] Cf. SCC, vol. III, pp. 133−137.

[49] Jerome Cardan, + 1501−1576, Italian mathematician who wrote also on medicine and the occult sciences.

[50] Ting Huan, inventor, mechanic, and artisan. Floruit ca. + 180. For information on ''Cardan'' suspension or Ting Huan's suspension, see SCC, vol. IV, part 2 (Cambridge University Press, 1965), pp. 228−236.

[a] 宋   [b] 元   [c] 丁緩   [d] 張衡

dynamics were relatively advanced but magnetic phenomena almost unknown. Yet China and Europe differed most profoundly perhaps in the great debate between continuity and discontinuity; just as Chinese mathematics was always algebraic rather than geometrical, so Chinese physics was faithful to a prototypic wave theory[51] and perennially averse to atoms. There is no doubt that the Buddhist philosophers were always bringing in knowledge of the Vaiśeshika theories about atoms, but nobody in China was willing to listen. The Chinese stuck to the ideas of universal motion in a continuous medium, action at a distance, and the wavelike motions of the Yin and Yang.

One most significant point is that although the Chinese of the Chou[a] and Han,[b] contemporary with the Greeks, probably did not rise to such heights as they, nevertheless in later centuries there was nothing at all in China corresponding to the Dark Ages in Europe. This fact is demonstrated well by the sciences of geography and cartography. Although the Chinese knew of discoidal cosmographic world maps, they were never dominated by them. Quantitative cartography began in China with Chang Hêng and P'ei Hsiu[c][52] at about the time that Ptolemy's[53] work was falling into oblivion in the West, indeed soon after his death, and it continued steadily with a consistent use of the rectangular grid right down to the coming of the Jesuits in the 17th century. Chinese geographers were also very advanced in the field of survey methods and the making of relief maps.

Mechanical engineering, and indeed engineering in general, were fields in which classical Chinese culture scored special triumphs. Both forms of efficient harness for equine animals, a problem of link-work essentially, originated in the Chinese culture area. There, too, water power was first used for industry, about the same time as in the West, in the + 1st or −1st century, not, however, for grinding cereals, but rather for the operation of metallurgical bellows. And that brings up something else, because the development of iron and steel technology in China constitutes a veritable

[51] Cf. SCC, vol. IV, part 1 (1962), pp. 9–10.

[52] P'ei Hsiu, + 229–271, notable Chinese cartographer and geographer.

[53] Ptolemy, + 2nd-century Alexandrian astronomer, produced works on geography and astronomy that dominated Western thought for the next twelve hundred years.

[a] 周   [b] 漢   [c] 裴秀

epic, with the mastery of iron casting occurring some fifteen centuries before its achievement in Europe. Contrary to the usual
ideas, mechanical clockwork began, not in early Renaissance
Europe but in T'ang[a] China, in spite of the highly agrarian
character of East Asian civilisation. Civil engineering also shows
many extraordinary achievements, notably iron-chain suspension
bridges, and the first of all segmental-arch bridges, the magnifi-
cent one built by Li Ch'un[b][54] in +610. Hydraulic engineering was
always prominent in China on account of the necessity to control
waterways and to develop river conservation, defence against floods
and drought, irrigation for agriculture, and the transport of tax
grain.

In martial technology the Chinese people also showed notable
inventiveness. The first appearance of gunpowder occurred in
China in the +9th century, and from +1000 onward there was a
vigorous development of explosive weapons some three centuries
before they were known in the West. The first appearance of a cannon in Europe is the bombard depicted in the Bodleian Library
manuscript of +1327, but you have to go back a good three centuries before that to see the beginning of the affair in China. We
know now that every single stage from the first development of the
gunpowder formula to the development of the iron-barrel cannon,
using the propellant force of gunpowder, was gone through in the
Chinese culture area before it ever came to the Arabs or to Europe
at all. Probably the key invention was that of the fire lance, the
huo ch'iang,[c] which we now know took place in the middle of the
+10th century. That was a device in which a rocket composition was
enclosed in a bamboo tube and used as a close-combat weapon.
From this thing derived, we have no doubt, all subsequent rockets,
barrel guns, and cannon of whatever material constructed.

Turning from the military to civilian, other aspects of technology
have great importance, especially that of silk, in which the Chinese
people excelled so early. Here the mastery of textile fibres of extremely long staple appears to have led to several fundamental
engineering inventions, for example, the first development in any
civilisation of the driving belt and the chain drive. It is also possi-

[54] Li Ch'un, bridge-builder, the first to use the segmental arch. Floruit ca.
+600.
[a] 唐   [b] 李春   [c] 火鎗

ble to say that the first appearance of the standard method of inter-conversion of rotary and longitudinal motion,[55] which found its great use in the early steam engines in Europe, came up also in connection with the metallurgical blowing-engines referred to already. If one is going in for epigrams, I ought to have mentioned when speaking of magnetism and of the magnetic compass that in China people were worrying about the nature of the declination (why the needle does not usually point exactly to the north) before Europeans had even heard about the polarity.

Nor was there any backwardness in the biological field either, because we find many agricultural inventions arising from an early time. There are texts which parallel those of the Romans, like Varro[56] and Columella[57] from a similar period; and one could take a very remarkable example from the history of biological plant protection. I do not know how many of us are conscious of the fact that the first case of insects' being set to destroy other insects, and so work in the service of man, occurred in China: in the *Nan Fang Ts'ao Mu Chuang*[a] written about + 340, there is a description of how the farmers in Kuangtung[b] and the southern provinces in general who grow oranges in groves go to the marketplace at the right time of the year and purchase little bags containing a particular kind of ant, which they then hang on the orange-trees. These ants completely keep down all the mites and spiders, and other insect pests, which would otherwise damage the orange crop. As a matter of fact, in China today very big things are going on in connection with biological plant protection. One of the experts in this field visited us in Cambridge the other day, and we could pick the *Nan Fang Ts'ao Mu Chuang* off the shelf and show her what her ancestors had done.

Medicine, again, is a field which it is rather absurd to bring up in a couple of minutes, because one could speak not for one hour, but many hours, on the subject of the history of medicine in China. It represents a field which aroused the intense interest of Chinese people all through the ages, and one which was developed by their special genius along lines perhaps more different from

[55] For details, see SCC, vol. IV, part 2, pp. 119–126.
[56] Varro, −116 to −27, learned and prolific Roman writer, especially on agriculture.
[57] Columella, + 1st-century Roman agricultural writer.
[a] 南方草木狀   [b] 廣東

those of Europe than in any other case. I think one might just refer, as an example, to the fact that people in China were free from the prejudices against mineral remedies which were so striking in the West. They needed no Paracelsus to awaken them from their Galenical slumbers, because they never participated in such slumbers; in other words, the Pên Ts'ao[a] books (the pharmaceutical natural histories) from the earliest times contained mineral and animal, as well as botanical, remedies. This was something that Europe did not have, because Galen[58] laid such emphasis on plant drugs that people were rather afraid to use minerals or any animal substance. And then of course there was the development of acupuncture and moxa, which are the subjects of a special lecture later in this series.

Coming now to the further examination of some of the great contrasts between China and Europe, I would like to stress that the *philosophia parennis* of China was an organic materialism. You can illustrate this from the pronouncements of philosophers and scientific thinkers of every epoch. Metaphysical idealism was never dominant in China, nor did the mechanical view of the world exist in Chinese thought. The organicist conception in which every phenomenon was connected with every other according to a hierarchical order was universal among Chinese thinkers. In some respects this philosophy of Nature may have helped the development of Chinese scientific thinking. For example, that the lode stone should point to the north, to the North Star, the pole star, the *pei chi,*[b] was not so surprising if one was already convinced that there was an organic wholeness in the cosmos itself. In other words, the Chinese were *a priori* inclined to field theories, and this might well account also for the fact that people in China arrived so early at correct ideas of the cause of the tides of the sea.[59] As early as the San Kuo[c] (Three Kingdoms, 220–265) period, one can find

---

[58] Galen, + 129–199, famous Greek physician in the Roman army service. Together with Hippocrates, he dominated Western medicine for fifteen hundred years.

[59] Until modern times there was, on the whole, more knowledge of, and interest in, the phenomena of tides in China than in Europe. Attempts were made at quite accurate observation of the flow and ebb of tides early in the + 11th century and there was a distinct notion of "influence" on the part of celestial bodies on the tides. For further details, see SCC, vol. III, pp. 483–494.

[a] 本草   [b] 北極   [c] 三國

remarkable statements of action at a distance taking place without any physical contact across vast distances of space.

We mentioned before that Chinese mathematical thought and practice were invariably algebraic, not geometrical. No deductive Euclidean geometry developed spontaneously in Chinese culture, and that was, no doubt, somewhat inhibitory to the advances the Chinese were able to make in optics—in which study, on the other hand, they were never handicapped by the rather absurd Greek idea that rays were sent forth by the eye. Euclidean geometry was probably brought to China in the Yüan period, but in did not take root until the arrival of the Jesuits. Nevertheless it is very remarkable that the lack of Euclidean geometry did not prevent the successful realisation of the great engineering inventions, including the highly complicated ones in which astronomical demonstrational and observational equipment was driven by water power through the use of elaborate gearing. Again, there was the interconversion of rotary and longitudinal motion already mentioned.

The clockwork story involved the invention of an escapement—in other words, a mechanical means of slowing down the revolutions of a set of wheels so that it would keep time with humanity's primary clock, the apparent diurnal revolution of the heavens. It is interesting that Chinese practice was not, as might seem at first sight, purely empirical, because the successful erection of the great clock tower of Su Sung[a] in K'aifêng[b] in +1088 was preceded by the elaboration of a special theoretical treatise by his assistant, Han Kung-Lien,[c] in which the trains of gears and general mechanics were worked out from first principles. He did not have Euclid, but he could do that. Something of the same kind had been done on the occasion of the first invention of the hydromechanical clock, by I-Hsing,[d][60] that great Tantric Buddhist monk, and Liang Ling-Tsan[e] early in the 8th century, six centuries before the first European mechanical clocks with their verge-and-foliot escapements. Moreover, although China had no Euclid, that did not prevent the Chinese from developing and consistently employing these astronomical coordinates which have completely con-

[60] I-Hsing, +682−727, Tantric monk, greatest mathematician and astronomer of his age, inventor, with Liang Ling-Tsan, of the water-wheel linkwork escapement.

[a] 蘇頌   [b] 開封   [c] 韓公廉   [d] 一行   [e] 梁令瓚

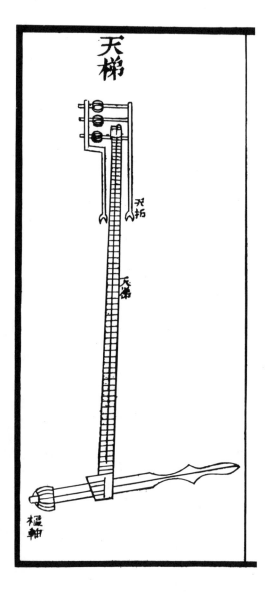

Fig. 1. The oldest chain-drive in any civilisation, part of Su Sung's astronomical clock, from *Hsin I Hsiang Fa Yao*, + 1094.

quered modern astronomy,[61] and are universally used today, nor did it prevent their consequent elaboration of the equatorial mounting even though there was nothing but a sighting tube, not as yet a telescope, to put into it.

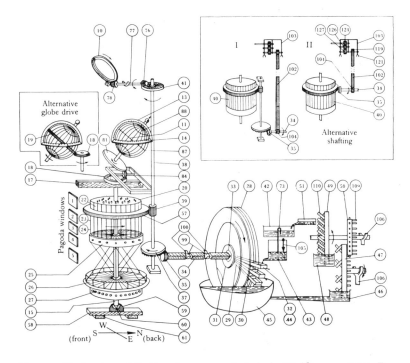

Fig. 2. Mechanism of Su Sung's astronomical clock, K'aifêng, + 1088. For details, see SCC, vol. IV, part 2, Fig. 652a.

I have mentioned already the wave-particle antithesis. The prototypic wave theory with which the Chinese concerned themselves from the Ch'in[a] and Han[b] onward was connected with the eternal rise and fall of the two basic natural principles, Yin[c] and Yang.[d] From the + 2nd century onward atomistic theories must have been introduced to China time after time, but they never took any root in Chinese scientific culture; yet this lack of particulate theory did

---

[61] For details, see SCC, vol. III, pp. 266–268.

[a] 秦   [b] 漢   [c] 陰   [d] 陽

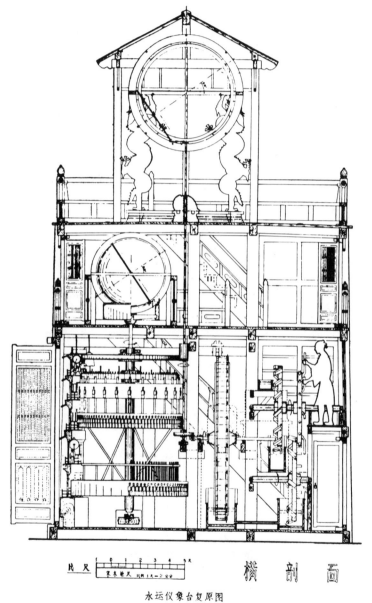

Fig. 3.   Reconstruction of Su Sung's astronomical clock-tower, K'aifêng, +1088, after Wang Chên-To, "Chieh K'ai Liao Wo Kuo 'T'ien Wên Chung' Te Mi-Mi," *Wên Wu Ts'an K'ao Tzu Liao*, 1958:9.

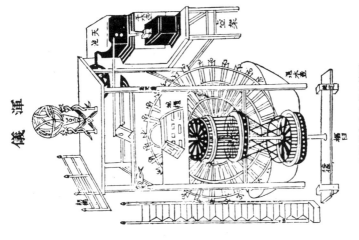

渾 儀

水運儀象台內部機械圖
（故守山閣叢刊本"新儀象法要"）

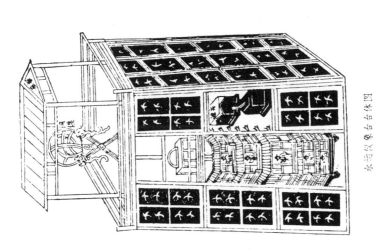

水運儀象台台体図
（故守山閣叢刊本"新儀象法要"）

Fig. 4. Su Sung's astronomical clock-tower, K'aifêng, +1088 (original illustrations from *Hsin I Hsiang Fa Yao*), after Wang Chên-To, "Chieh K'ai Liao Wo Kuo 'T'ien Wên Chung' Te Mi-Mi."

not prevent the Chinese from curious achievements such as the recognition of the hexagonal system of snowflake crystals many centuries before this was first noticed in the West. Nor did it hinder them from helping to lay the foundations of the knowledge of chemical affinity, as was done in some of the alchemical tractates of the T'ang, Sung, and Yüan. There the absence of particulate conceptions was probably less inhibitory than it otherwise might have been, because it was, after all, only in the post-Renaissance period in Europe that such theories became so fundamental for the rise of modern chemistry.

Fig. 5. Reconstruction of Su Sung's astronomical clock-tower, K'aifêng, + 1088, by John Christiansen.

Fig. 6.   Model of Su Sung's astronomical clock-tower, by John Combridge, in the Science Museum, South Kensington, London.

I should not want to quarrel altogether with the idea that the Chinese were a fundamentally practical people, inclined to distrust all theories. One must beware, however, of carrying that idea too far, because the Neo-Confucian school in the +11th, 12th, and 13th centuries achieved a wonderful philosophical synthesis, strangely parallel in time with the scholastic synthesis of Europe. One might almost say that the disinclination of the Chinese to engage too far in theory, especially geometrical theory, brought advantages with it. For example, Chinese astronomers did not reason about the heavens, like Eudoxus[62] or Ptolemy, but they did avoid the assumption that the heavens were composed of concentric crystalline solid spheres, an idea which completely dominated mediaeval Europe. By a strange paradox, when Matteo Ricci,[63] the great leader of the Jesuit Mission, came to China at the end of the +16th century, he mentioned in one of his letters home a number of the foolish ideas entertained by the Chinese, among which prominently figured the fact that they did not believe in crystalline solid celestial spheres; but it was not long before the Europeans gave up the idea themselves.

This fundamental practicality did not imply an easily satisfied mind, because very careful experimentation was practised in classical Chinese culture. The discovery of magnetic declination[64] would never have occurred unless the geomancers had been attending most carefully to the positions of their needles; and the triumphs of the ceramics industry could never have been achieved without some fairly accurate form of temperature measurement and control, and the reproduction at will of oxidising and reducing conditions inside the kilns. The fact that relatively few of these technical details have come down to us springs from social factors which prevented the publication of the records which the higher artisans most certainly kept, though we do get instances of such records from time to time: for example, there was the *Mu Ching*[a] (The Timberwork Manual) which was the basis of the *Ying Tsao Fa Shih*,[b] the great classical work on architecture of +1102. The *Mu*

---

[62] Eudoxus, −404 to −352, eminent Greek astronomer and geometer.

[63] Matteo Ricci, +1552−1610, Italian missionary, first head of the Jesuit Mission, introduced into China contemporary European developments in mathematics, astronomy, and other sciences.

[64] Cf. SCC, vol. IV, part 1, pp. 293−312.

[a] 木經   [b] 營造法式

Ching was the work of a famous pagoda architect, Yü Hao,[a] but it must have been dictated by him, because although he certainly could not read or write he was able to pass on his information. Another example is the famous *Fukien*[b] *Shipbuilders' Manual*, a rather rare manuscript which shows that the artisans had friends who could write, and who could use the technical terms, and who wrote down in books what the artisans were able to tell them.

Here we are brought up against the social and economic questions about which I would like to talk in the last few minutes of this lecture, because they are exceedingly important for the comparative study of Chinese and Occidental science, technology, and medicine. It is probably impossible to understand the situation without realising the enormous differences in social and economic structure between traditional China and the traditional West. Although many differences of interpretation exist among scholars, I feel quite satisfied with the broad principle that during approximately the past two thousand years, China did not have feudalism in the aristocratic military Western sense. Whether the Chinese system is known, as it was by the founding fathers of Marxism, as the "Asiatic mode of production," or (as other people have called it) "Asiatic bureaucratism," or "feudal bureaucratism," or (as Chinese friends very often used to like to call it when I was in China during the war) "bureaucratic feudalism," or whatever other term you like to adopt, it was certainly something different from what Europe knew.

I have sometimes been tempted to regard it as a disappearance of all the intermediate feudal lords at an early stage of the unification of the empire, after the time of Ch'in Shih Huang Ti[c] in the −3rd century. The country was ruled by only one feudal lord, namely the emperor himself, operating and exploiting by means of a relatively hypertrophied instrument, the non-hereditary civil service, the bureaucracy or mandarinate, recruited from the scholar-gentry. It is very doubtful to what extent they should be called a "class" because it is clear that in different times and different degrees it had a great deal of fluidity. Families rose into the "estate," if you like, of the scholar-gentry, and sank out of it again, especially during those periods when the imperial examinations played such an important part in the recruiting of the civil service;

[a] 喻皓   [b] 福建   [c] 秦始皇帝

families which could not produce the right talents and the particular skills and gifts for success in the examinations and the
bureaucratic service were not likely to survive at a high social level
for more than a generation or two. Thus the *shih*, the scholarbureaucrats, were the literary and managerial elite of the nation for
two thousand years, and we must not forget that the conception of
the *carrière ouverte aux talents*, the "career open to talent," which
many people date from the French Revolution, was neither French
nor even European; it had been Chinese for a millennium already.
You can even show by chapter and verse that the theory of competitive examination for the civil service was taken over by the
Western nations in the 19th century in full consciousness of the
Chinese example, even though the sinophilism of the chinoiserie
period had given place some time before to a certain disenchantment regarding the celestial empire and its mandarinate as a college of all sages. Of course the mandarinate was not as classless as
has sometimes been made out, for, even in the best and most open
periods, boys from learned homes which had good private libraries
had a great advantage; but in any case it remains a fact that the
scale of values of the scholarly administrator differed profoundly in
all ages from that of the acquisitive merchant.

How this affected science and technology is a very interesting
and complicated question, into which we cannot go far for lack of
time, but there is no doubt that in China certain sciences were orthodox from the point of view of the scholar-gentry, and others
not. The institution of the calendar, and its importance for a
primarily agrarian society, and also to a lesser extent the belief in
state astrology, made astronomy always one of the orthodox
sciences. Mathematics was considered suitable as a pursuit for the
educated scholar, and so also physics up to a point, especially as
both mathematics and physics contributed to the engineering
works so characteristic of the centralised bureaucracy. The need of
Chinese bureaucratic society for great works of irrigation and water
conservation meant not only that hydraulic engineering was
regarded favourably among the traditional scholars, but also that it
helped in its turn to stabilise and support that form of society of
which they themselves were such an essential part. Many people
have believed that the origin and development of feudal
bureaucratic society in China was at least partly dependent on the
fact that from very early times the undertaking of great hydraulic

engineering works tended to cut across the boundaries of the lands of the individual feudal lords, and this had the effect of concentrating all power in the centralised bureaucratic imperial state. As a matter of fact, you can find statements in certain texts, such as the *Yen T'ieh Lun*[a] (Discourses on Salt and Iron) which dates from −81. This contains a page which says that the Son of Heaven has to consider the hydraulic engineering needs of vast areas, far more than any individual feudal lord has to worry about.

In contrast with these forms of applied science, alchemy was distinctly unorthodox, the characteristic pursuit of disinterested Taoists and other recluses. Medicine was in this respect rather neutral; on the one hand the demands of traditional filial piety made it a respectable study for the scholars—indeed it was more and more entered by the *Ju i*,[b] the Confucian physicians—while on the other hand its necessary association with pharmacy connected it with the Taoists, alchemists, and herbalists.

In the end, I believe we shall find that the centralised feudal bureaucratic style of social order was in the early stages favourable to the growth of applied science. Take the case of the seismograph, which I have already mentioned more than once. It is parallelled by the existence of rain gauges and even snow gauges at a remarkably early time, and it is highly probable that the stimulus of such inventions came from the very reasonable desire of the centralised bureaucracy to be able to foresee the shape of things to come. Thus, for example, if a particular region was hit by a severe earthquake, it would be advisable to know this as soon as possible, in order that help might be sent and reinforcements supplied to the local authorities in case of a popular uprising. Similarly, the rain gauges on the edge of the Tibetan massif would have played a useful part in determining the measures to be taken for the protection of the hydraulic engineering works lower down. Moreover, Chinese society in the Middle Ages was able to mount much greater expeditions and much more organised scientific field work than was the case in any other society of that time. A good example of this is the survey of the meridian arc carried out early in the + 8th century under the auspices of I-Hsing,[c] whom I mentioned before, and the astronomer Nankung Yüeh.[d][65] This was a geodetic

[65] Nankung Yüeh, Astronomer-Royal, responsible, with I-Hsing, for the early + 8th century meridian arc survey. Floruit ca. + 700.

[a] 鹽鐵論   [b] 儒醫   [c] 一行   [d] 南宮說

survey covering a line no less than 2500 km long, ranging from Indo-China to the borders of Mongolia. At about the same time, an expedition was sent down to the East Indies for the purpose of surveying the constellations of the southern hemisphere within 20° of the south celestial pole. It is very doubtful whether any other state in the world at that time could have successfully engaged in such far-flung activities.

From early times Chinese astronomy had benefitted from state support, but the semi-secrecy which it involved was to some extent a disadvantage. Chinese historians sometimes realised this; for example, in the dynastic history of the Chin,[a] in the *Chin Shu*,[b] there is an interesting passage which says that "astronomical instruments have been in use from very ancient days, handed down from one dynasty to another, and closely guarded by official astronomers. Scholars therefore have had little opportunity to examine them, and this is the reason why unorthodox cosmological theories have been able to spread and flourish so much." However, you cannot push that argument too far. It is clear that in the Sung[c] period, at any rate, the study of astronomy was quite possible and even usual in scholarly families connected with the bureaucracy. We know, for example, that Su Sung[d] in his early years had model armillary spheres of small size in his home, and so gradually came to understand astronomical principles. About a century later the great philosopher Chu Hsi[e][66] also had an armillary sphere in his home and tried hard to reconstruct the waterpower clock-drive of the Sung, though unsuccessfully. Furthermore, there were periods, for example in the 11th century, when mathematics and astronomy played a prominent part in the famous official examination for the civil service.

[66] Chu Hsi, +1130−1200, greatest philosopher in Chinese history, climax of Neo-Confucianism.

a 晉   b 晉書   c 宋   d 蘇頌   e 朱熹

2

The Epic of Gunpowder and Firearms, Developing from Alchemy

The development of gunpowder and gunpowder weapons was certainly one of the greatest achievements of the mediaeval Chinese world. One finds the beginning of it toward the end of the T'ang,[a] in the +9th century, when the first reference to the mixing of charcoal, saltpetre (potassium nitrate), and sulphur is found. This occurs in a Taoist book which strongly recommends alchemists not to mix these substances, especially with the addition of arsenic, because some of those who have done so have had the mixture deflagrate, singe their beards, and burn down the building in which they were working.

Let us go back to some of the earliest experiments of which we have records which led to the invention of the gunpowder formula. In the first place, the ancient Chinese were very adept at the making of smoke, the burning of incense, and fumigation as such. This procedure was carried on for hygienic and insecticidal reasons, and it is found even in the *Shih Ching*,[b] where the annual purification of dwellings, a New Fire ceremony, and so on are mentioned. The *Kuan Tzŭ*[c] Book, not many centuries later, refers to the medicinal fumigation of houses, closing all the apertures, and we know that insecticidal plants like *Illicium* and *Pyrethrum* were used in those operations. Then we know also how from Ch'in[d] and Han[e] times onward Chinese scholars fumigated their libraries to keep down the depredations of bookworms.

The Chinese were really great smoke-producers. Toxic smokes and smoke-screens generated by pumps and furnaces in siege warfare are referred to in the military sections of the *Mo Tzŭ*[f] book, dating from the -4th century. There may be sources earlier than that but we do not know them. The *Mo Tzŭ* book is certainly full

[a] 唐　　[b] 詩經　　[c] 管子　　[d] 秦　　[e] 漢　　[f] 墨子

of such toxic devices, and they prefigure the toxic smoke-bombs in
the *Huo Lung Ching*[a] of the + 15th century, which come also in
the *Wu Ching Tsung Yao*[b] of + 1044, which I shall quote below.
This was a wonderful book, a "Compendium of the Most Important Military Techniques" compiled by Tsêng Kung-Liang[c1] during the Northern Sung period. The sea-battles of the + 12th century between the Sung[d] and the Chin[e] Tartars, as well as the civil
wars and rebellions of the time, show many other examples of the
use of toxic smokes containing lime and arsenic. Indeed, the earth-
shaking invention, literally earth-shaking, of gunpowder itself
sometime in the + 9th century, as I've said, was closely related,
because it derived from incendiary preparation and its earliest formulae sometimes contained arsenic.

As always, of course, good things developed as well as bad
things. For example, in + 980 the monk Tsan-Ning[f2] wrote in his
Ko Wu Ts'u T'an[g] (Simple Discourses on the Investigation of
Things) that "when there is an epidemic of febrile disease, let the
clothes of the sick persons be collected as soon as possible after the
onset of the malady, and thoroughly steamed. In this way the rest
of the family will escape infection." Now, that would have in-
trigued Louis Pasteur[3] and Joseph Lister.[4] The evil and beneficent
effects of knowledge have always gone hand in hand, for such is
man's nature.

Another important point, of course, was the early recognition
of saltpetre, potassium nitrate. Until that was thoroughly
understood, and the salt could be separated and crystallised, it was
no good expecting the appearance of gunpowder. There is an interesting book, the *Chu Chia Shên P'in Tan Fa*,[h] in the *Tao
Tsang*,[i] which gives much information about these things. Another

---

[1] Tsêng Kung-Liang, + 998—1078, military encyclopaedist, whose book of
1044, *Wu Ching Tsung Yao*, gave the first gunpowder formulae in any civilisation.
[2] (Lu) Tsan-Ning, + 919—1001, Buddhist monk, scientist, chemist, and
microbiologist.
[3] Louis Pasteur, 1822—1895, scientist, chemist, and microbiologist, founder of
the science of bacteriology.
[4] Joseph Lister, 1827—1912, English surgeon, introducer of antisepsis.

[a] 火龍經   [b] 武經總要   [c] 曾公亮   [d] 宋   [e] 金
[f] 贊寧   [g] 格物麤談   [h] 諸家神品丹法   [i] 道藏

story in a related book, the *Chin Shih Pu Wu Chiu Shu Chüeh*[a] (Explanation of the Inventory of Metals and Minerals according to the Numbers Five and Nine), mentions the appearance of Sogdian Buddhist monks who knew about saltpetre in the + 6th century and noticed its presence as an incrustation on the soil. There is an interesting quotation from the Lin-Te[b] period of the T'ang (+ 664), when a certain Sogdian[5] monk called Chih Fa-Lin[c6] came to China bringing with him some sutras in the Sanskrit language for translation.

> When he reached the Ling-Shih[d] district in Fên-chou,[e] he said, "This place must be full of saltpetre, why isn't it collected and put to use?" At that time he was in the company of twelve persons, and together they collected some of the substance and tested it but found it unsuitable for use, and not comparable with that produced in Wu-Ch'ang.[f] Later they came to Tsê-chou[g] and the monk said again that saltpetre must also occur in this region: "I wonder if it will be as useless as what we came across before?" Whereupon they collected the substance, and on burning it emitted copious purple flames. The Sogdian monk said, "This is a marvellous substance which can produce changes in the five metals, and when the various minerals are brought into contact with it they are completely transmuted into liquid form." And in fact its properties were indeed the same as the material from Wu-Ch'ang which they knew about already.

So here you have a mention of the potassium flame, the use of saltpetre as a flux in smelting, and its ability to liberate nitric acid, which would help the solution of inorganic substances hard to dissolve.

In the *Chu Chia Shên P'in Tan Fa* that I mentioned just now, there is an interesting account of experiments which may have been made by the great alchemist and physician Sun Szŭ-Mo[h7] in about + 600. One of the formulae says:

> Take sulphur and saltpetre two ounces each, grind them together, and then put them in a silver-melting crucible or refractory pot. Dig a pit in the ground and put the vessel inside it so that its top is level with the surface

---

[5] Sogdian. Sogdiana is the ancient region corresponding to more recent Bokhara. It is bounded by the Oxus River to the south, and by the Jaxartes to the north.

[6] Chih Fa-Lin, monk from Central Asia. Floruit ca. + 664.

[7] Sun Szŭ-Mo, + 581–672, eminent Sui and T'ang alchemist, wrote the *Tan Ching Yao Chüeh*.

[a] 金石簿五九數訣   [b] 麟德   [c] 支法林   [d] 靈石   [e] 汾州

[f] 烏長   [g] 澤州   [h] 孫思邈

and cover it all round with earth. Take three perfect pods of the soap-bean tree, uneaten by insects, char them so that they keep their shape, and then put them in the pot with the sulphur and the saltpetre. After the flames have subsided, close the mouth and place three catties of glowing charcoal on the lid, and when this has been consumed, remove it all. The substance need not be cool before it is taken out; it has been subdued by fire.

Someone seems to have been engaged here, perhaps round about + 650, in an operation designed, as it were, to produce potassium sulphate, and not therefore very exciting; but on the way he stumbled upon the first preparation of a deflagrating, and later explosive, mixture in the history of all civilisation. Exciting must have been the word for that. But of course he may not have realised quite what he was doing and what had really happened.

Then we have an interesting book by Chao Nai-An,[a8] possibly of about + 808, or a bit later, a florilegium of chemical writings in five chapters. There he has an experiment which takes its place as another of the earliest known records of a protogunpowder mixture, described under the heading of *fu huo fan fa*,[b] or a method of subduing alum or vitriol by fire, with a composition of sulphur, saltpetre, and dried *Aristolochia* as the carbon source. This would have ignited suddenly, bursting into flames without actually exploding. The exact sequence of these first accounts has yet to be determined, but if Sun Szŭ-Mo was really the experimenter of the *Chu Chia shên P'in Tan Fa*,[c] the middle of the + 7th century would have seen that first beginning; and it does look like the most archaic procedure, for the carbon source in the shape of the soap-bean pods was undoubtedly added with a very different intention.

Finally, among these early references, I would like to mention that interesting book, the *Chên Yüan Miao Tao Yao Lüeh*[d] (Classified Essentials of the Mysterious Tao of the True Origin of Things). We do not know its exact date but it was probably about the middle of the 9th century. This is the book referred to above because it mentions no less than thirty-five different elixir formulae which the writer points out to be wrong or dangerous, though popular in his time. It tells of cases where people died after consuming elixirs prepared from mercury, lead, and silver; and of other cases where people suffered from boils or sores on the back

[8] Chao Nai-An, T'ang alchemist. Floruit ca. + 800.

[a] 趙耐庵   [b] 伏火礬法   [c] 諸家神品丹法   [d] 眞元妙道要略

after ingesting cinnabar; and of serious illness when people drank
"black lead juice," possibly a hot suspension of graphite.
Among the erroneous methods are boiling the ash obtained from
burning mulberry wood and regarding it as *ch'iu chih*[a] (autumn
mineral), or mixing common salt, ammonium chloride, and urine,
evaporating to dryness and calling the sublimate from that *ch'ien
hung*[b] (literally, lead and mercury). These look like falsifications
intended to deceive. Finally, among all these methods which the
writer warned were misleading and wrong, it says quite clearly that
some of the alchemists had heated sulphur together with realgar
(arsenic sulphide), saltpetre, and honey, with the result that their
hands and faces had been scorched when the mixture deflagrated,
and even their houses burned down. These things only bring
Taoism into discredit, he claims, and alchemists should not do
them. This passage is of outstanding importance because it is one
of the first references in any civilisation to a deflagrative or explosive mixture, protogunpowder, combining sulphur with nitrate
and a source of carbon.

After that, things started to happen quite rapidly. *Huo yao*[c] was
the common term for gunpowder in Chinese culture, the "fire
chemical," and as we hardly ever meet with the term in any other
context, its use is a sure indication that gunpowder is being talked
about. There is one exception to that, namely its use in *nei tan*,[d]
inner alchemy, or what we call physiological alchemy, where it can
have another significance; but generally speaking, it always refers
to a gunpowder mixture of one kind or another. We meet with the
first use of *huo yao* as a slow match for a flame-thrower in + 919,
and by the time we reach the year + 1000 the practice of using
gunpowder in simple bombs and grenades was coming into use,
especially thrown or lobbed over from trebuchets, which got the
name of *huo p'ao*.[e]

Here the general chart shows the chronological order in which
these different developments took place. The flame-thrower which
used gunpowder as a slow match was indeed a fascinating piece of
machinery. It is described and illustrated in the *Wu Ching Tsung
Yao*[a] of + 1044, and it was a naphtha projector like the "syphon"
of the Byzantine Greeks. Actually, it was a very interesting force-

[a] 秋石   [b] 鉛汞   [c] 火藥   [d] 內丹   [e] 火砲
[f] 武經總要

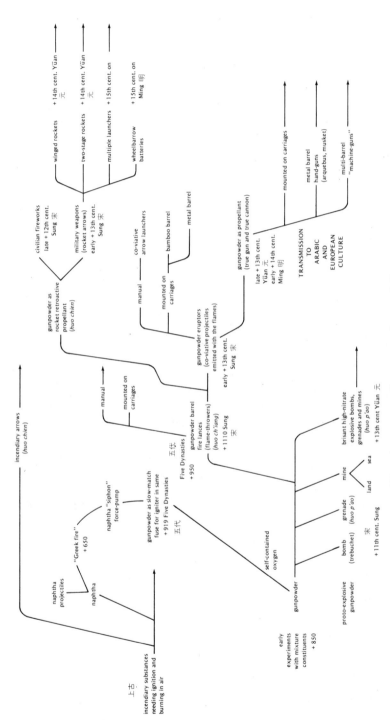

Development of gunpowder and firearms

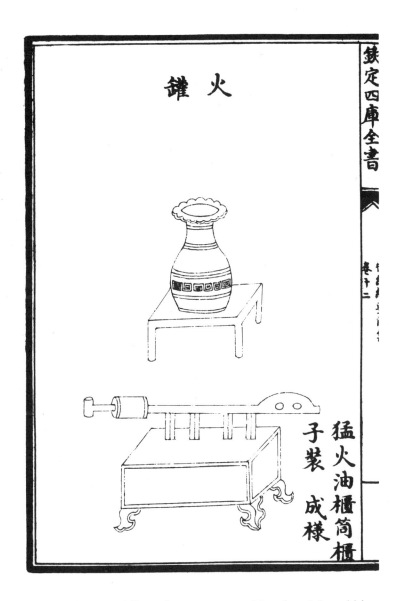

Fig. 7.　Below, petrol flame-thrower pump and jar of naphtha, which text says is good for firing floating bridges, from *Wu Ching Tsung Yao* (+ 1044), *Ch'ien*, Ch. 12.

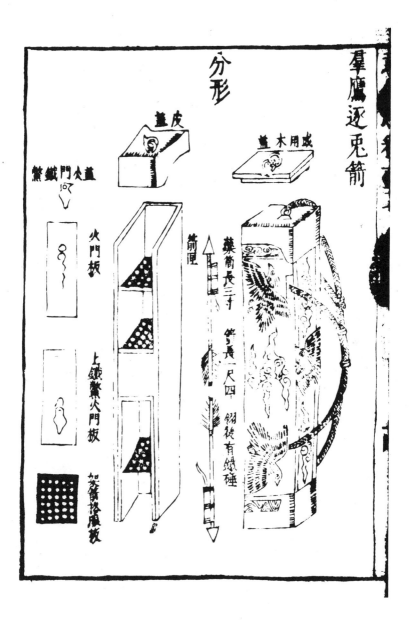

Fig. 8.   Hawk flock, rocket-arrow launcher, from *Wu Pei Chih* (+ 1628), Ch. 127.

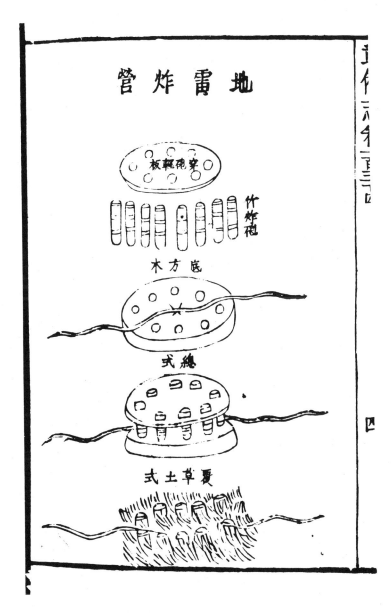

Fig. 9. Land mine, again formed of individual fire lance tubes, from *Wu Pei Chih*, Ch. 134.

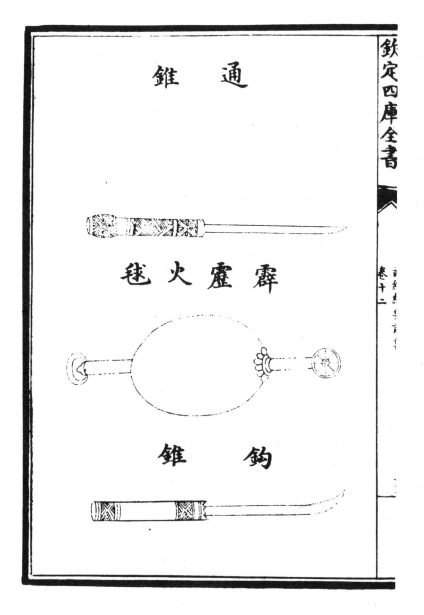

Fig. 10. *P'ili p'ao* weak-casing bomb, poison smoke bomb or bamboo grenade, from *Wu Ching Tsung Yao*, *Ch'ien*, Ch. 12.

pump because it had two pistons on one piston rod, and it drew up from the tank below the naphtha or low-boiling-point petroleum fractions, then ignited the liquid and shot it forth for many yards. It must have been quite frightening for anyone trying to climb over a city wall.

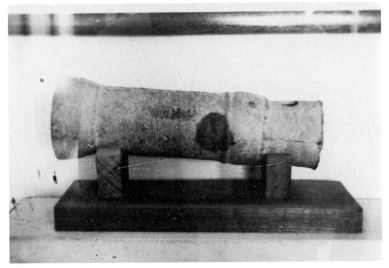

Fig. 11.   Early bombard (bronze), dated +1332, National History Museum, Peking.

The first formula for the composition also appeared in the *Wu Ching Tsung Yao* compiled by Tsêng Kung-Liang[a] in +1044—a great deal earlier than the first appearances or references to any gunpowder composition in Europe. For those you have to go forward to +1327, contemporary with Mongol times, at best 1285. That is an important date to keep in mind because it was the first occurrence of any reference to the gunpowder formula in Occidental civilisation.

Of course, the bombs and grenades of the first part of the 11th century did not contain a brisant explosive like that which became known in the following two centuries when the proportion of nitrate was raised. At first the proportion of nitrate, which is the substance that provides the oxygen in the mixture, was low, but

[a] 曾公亮

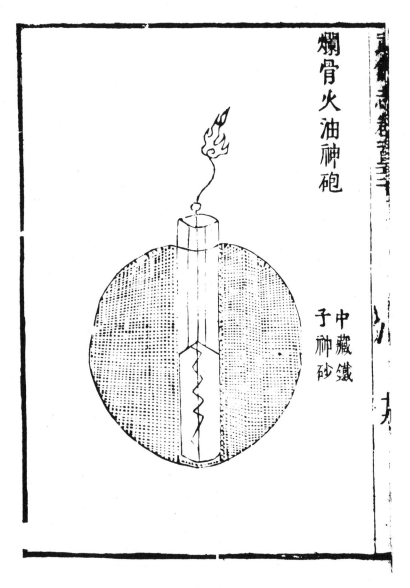

Fig. 12. Bomb containing gunpowder at centre, missiles, poison smoke and lachrymatory smoke, *tung* oil 桐油, urine solids 銀銹, sal ammoniac 硇砂, faeces 金汁, garlic juice 蒜汁, roasted iron filings 炒製鐵砂, and broken porcelain 磁粉 in a cast-iron case, from *Wu Pei Chih*, Ch. 122.

later on it was raised. These early forms of protogunpowder were more like rocket compositions, going off with a "whoosh." That could be quite frightening, but it was not a destructive explosion. By the middle of the 13th century, when the Mongols and the Sung[a] Chinese were locked in combat, the proportion of nitrate was at last raised to the point where really destructive explosions could take place, walls could be blown up, and city gates broken in.

Fig. 13.   The oldest representation of a fire lance, Buddhist painted banner from Tunhuang, ca. + 950.

These followed the very important transition to the barrel gun. It occurred, we now think, in the middle of the + 10th century, in other words in the Wu Tai[b] (Five Dynasties) period, when the fire lance first came into existence—that's the *huo ch'iang*,[c] the fire lance or fire spear. A most remarkable banner from Tunhuang[d] has been discovered recently in the Musée Guimet in Paris. It shows the Buddha meditating, and the hosts of Mara the Tempter all round the side, looking very fierce and throwing things at him. Many of them are dressed in military uniforms, and in one place there is a

[a] 宋   [b] 五代   [c] 火鎗   [d] 敦煌

figure with three snakes in its headdress—a devil, in fact—holding
in its hands a cylinder from which are issuing flames. The fact that
flames are not going upward but are being shot forth horizontally
is a clear indication that the object can be nothing but a fire lance.
It must contain rocket composition, which is bursting forth like a
miniature three-minute flame-thrower, producing the effect
desired.

　　It is easy to see from this the great importance of the availability
of a natural form of tubing, the stem of the bamboo; and we
would like to maintain that this was in fact the ancestor, the
original ancestor, of all barrel guns and cannon of every kind. The
fire lance played a very prominent part in the wars between the
Sung and the Jurchen Chin[a] Tartars from about +1100 onward.
There exists, for example, a book by Ch'ên Kuei[b] called the *Shou
Ch'êng Lu,*[c] a record of the defence of a city north of Hankow[d]
about +1120, in which is described the use of many of these *huo
ch'iang* tubes filled with rocket composition and held upon the
end of a spear. In my opinion, an adequate supply of these three-
minute flame-throwers, passed on from hand to hand, must have
effectively discouraged enemy troops from storming one's city
wall.

　　By about +1230 we begin to have descriptions of really destructive explosions, as I said before, in the later campaigns between the
Sung and the Yüan[e] Mongols; and then, about 1280, comes the
appearance of the true metal-barrel gun or cannon somewhere in
the Old World. There has been a great deal of controversy and
doubt as to where it first appeared, whether among the Arabs with
their *madfaa,* as they were called, or possibly among the
Westerners. Between +1280 and +1320 is the key period, the
crucial period, for the appearance of metal-barrel cannon, but we
have no doubt whatever that its real ancestry was the substantial
bamboo tube of the Chinese fire lance.

　　We have to follow this a bit further through several developments of great significance before we can talk about other important inventions connected with gunpowder. I would like first of all
to point out how easy and logical was the development of the fire
lance from the flame-thrower, the "fierce fire oil machine," the
mêng huo yu chi,[f] using "Greek fire" (that is, naphtha) or dis-

---

[a] 金　　[b] 陳規　　[c] 守城錄　　[d] 漢口　　[e] 元　　[f] 猛火油機

tilled light petroleum fractions of low boiling point. First it turned out that the petrol-projector pump could be made into a portable hand-weapon flame-thrower. Second, gunpowder, even though very low in nitrate, had already been used in that force-pump as a slow magic igniter, so the transition was very easy. It is interesting to note that Greek fire itself goes back to a chemist named Collinicus in 7th-century Byzantium, and that naphtha was used freely in the wars of the Arabs, while by the 10th century the rulers of the Wu Tai period in China were often giving presents of it to each other. So much was being passed around in those days that the Chinese must have been distilling it themselves.

The fire lance lasted in use until quite recent times. There still exists a photograph taken on a pirate ship in the South China Seas about sixty or seventy years ago, which shows the fire lance in action. It was well calculated to set the rigging or the woodwork of another ship on fire, and it was employed right down to the beginning of the present century.

I said just now that the amounts of petrol, or naphtha (low-boiling-point petroleum fractions) were so great in the Wu Tai period that the Chinese must have been distilling it themselves. It cannot all have been imported from Arabic lands. There were three ancient types of still. First, the Hellenistic still collected the distillate in a peripheral annular rim whence it flowed out by a side tube. Second, there was the Indian type of still, the Gandharan still, also with no cooling, where only vapour came over, and the distillate collected in the receiver. This was intended for mercury, but light petroleum fractions could have been distilled in this way; they could also have been distilled in the third type, the characteristic Chinese type of still. Here there was always a cooling vessel at the still-head, and a collecting bowl underneath it, with a side tube taking the distillate off to the receiver.

We have seen that the fire lance, the *huo ch'iang*, was certainly in existence by +950 and very prominent by +1110. Of course, as I have said, the gunpowder which it contained was not a brisant explosive, but more like a rocket composition, deflagrating violently and shooting forth powerful flames, but not going off suddenly with a mighty bang. At first, fire lances were held manually by the fire-weapon soldiers, but by the time of the Southern Sung they were made of bamboo much larger in diameter, perhaps up to a foot across, and mounted on a framework of legs, sometimes even provided with wheels so as to make them moderately mobile. This gave rise to a weapon for which we have found it necessary to coin a word; and we call it an "eruptor," because nothing or almost nothing like it existed in the Western world. (There are one or two exceptions; for example, something of the same kind was trundled out by the defenders of Malta in the siege against the Turks in + 1563, but they had no convenient name, and in our opinion they betrayed, together with many other things, their direct indebtedness to Chinese origins.)

Even more remarkable, these eruptors were so constructed as to shoot out projectiles along with the flames. Once again we need a new word for this, and we have decided to call these objects "co-viative" projectiles. They could be just bits of old iron, or even broken pottery or glass, but this system was quite different from the chain shot of later Napoleonic Europe, because there the function of the gunpowder was explosively propellant, and the chain shot took the place of the normal solid cannonball. The co-viative projectiles of the eruptors of the Sung and Yüan were more like case shot, which Mainwaring[9] in + 1644 defined as "any kind of old iron, stones, musket-bullets or the like which we put into cases to shoot forth out of our great ordnance." But the difference was that in the older Chinese system the pieces of hard, sharp-edged rubbish were actually mixed with the rocket composition, the gunpowder. Other names for the case shot in later times were cannister shot and langrel, but none of these things were co-viative, since that belonged to a much earlier stage of the story. Generally the eruptors consisted of bamboo barrels mounted on carriages, but it was precisely in connection with these that the first metal barrels appeared, cast in bronze or iron, a most important event. It is most notable that metal-barrel eruptors preceded metal-barrel bombards and cannon.

One extraordinary fact is that before the end of the eruptor period, actual explosive shells were fired forth as co-viative projectiles, and that must have been the time of their first invention. But eruptors with co-viative projectiles could be made small enough to be held manually, and by the late + 13th and early + 14th century, when all this was in its prime, co-viative arrow-launchers were also used. The arrows probably did not fly very far, since the gun-

[9] Charles Mainwaring, + 17th-century English gunner.

powder was not exerting its full propellant force, but for close combat on city walls their effects may have been impressive enough, especially against personnel armoured lightly or not at all. Pictures in late books show co-viative projectiles in hand-held eruptors or fire lances in their later form.

Lastly, there appeared the metal-barrel firearm characterised by two other basic features: first the use of high-nitrate gunpowder, and second the total occlusion of the muzzle (or front orifice) by a projectile (such as a bullet or cannonball) in such a way that the gunpowder exerted its full propellant effect. This type of firearm may be described as the true gun or cannon, and if it appeared in early Yüan times about 1280, as we suspect it did, its development had taken just about three and a half centuries since the invention of the first of the firearms, flame-throwers.

The bombard (as it may now be called) made its first appearance in Europe in + 1327, as we know from the famous manuscript in the Bodleian Library at Oxford. We must not imagine that at this early time there was a long, smooth bore with parallel walls to guide the projectile to its destination. The first bombards of Europe were distinctively vase-shaped with a rounded belly and a muzzle that splayed outward like the mouth of a blunderbuss. The shooting, therefore, must have been very hit-or-miss, but presumably the charge of gunpowder was rammed down into the bombard and the ball packed into the narrowest part, and then, even if the gunners (bombardiers) could not aim accurately at anything, their bombards would have been quite useful against castle walls or city gates or the massed troops of men in close order that probably moved about in those days.

Now, the interesting thing is that we find Chinese drawings of just such bombards. Pictures exist of a whole set of them, mounted on a carriage, exactly similar in shape to the first European 14th-century ones. So the probability is that they originated in China and were copied exactly in the West, where the beginnings of the knowledge of gunpowder itself go back only to 1285 or so. If this probability is correct, it would mean that the purely propellant phase of gunpowder and shot, the culminating stage of all the gunpowder uses, was attained in China with these bottle-shaped bombards before any knowledge of gunpowder itself reached Europe at all. Or perhaps it was about the same time. In any case, the whole development, beginning with the first experiments of

Sun Szŭ-Mo[a] and his friends, would have taken just on seven centuries—which was not bad going for the Middle Ages.

Here it is important to note, too, that archaeological finds of bronze and iron bombards and cannons in China have revealed many self-dated specimens, all much older than any of those that have been found in Europe. I am not sure that the inscriptions have yet taken us as far as + 1327 or beyond, but there are certainly many from the decades immediately following—and nothing as old can be demonstrated in Europe.

Now, the bombards with metal barrels were generally mounted on gun carriages, and it was not long before they were reduced in size to form guns which could be carried and fired by a single person, whence the line ran straight to the arquebus and the musket. Later on, in the + 16th century, the Chinese were deeply impressed by the muskets of the Portuguese, which they called *fo lang chi*[b] (Frankish devices), but that's another story we have no time to go into here. Chinese people were also impressed by the Portuguese light swivelling shipboard cannons or breech-loading culverins with removable metal cartridge holders, which they called *niao tsui chi*[c] (bird-beak guns), but again that falls outside the crucial period we are discussing at present.

Finally, long before that, both the bombards and the guns were mounted on stands in multiple batteries. The difficulty of knowing whether the vase-shaped bombards first appeared in China or in Europe arises largely from the peculiarities of the literature at both ends of the Old World. The Western chroniclers do not provide very much information until a rather later date, so that the iconographic evidence has particular importance, whereas in China we are faced with the difficulty that the technical books come at rather widely spaced intervals and in several different editions which differ among themselves, and are not always very precisely datable.

We have already mentioned the *Wu Ching Tsung Yao*[d] assembled by Tsêng Kung-Liang[e] in + 1044. I once found a Ming[f] edition of this in the Liu-Li Ch'ang[g] in Peking from which the whole of the gunpowder chapter was missing, so the information of that time was evidently still "restricted," and eventually I presented it to the

a 孫思邈   b 佛郎機   c 鳥嘴機   d 武經總要   e 曾公亮
f 明   g 琉璃廠

Library of Academia Sinica. The next landmark was the *Huo Lung Ching*[a] (The Fire-Drake Manual) which comes in half a dozen different parts and versions, associated with a variety of authors' names, some evidently fictitious, such as Chuko Liang,[b10] others quite likely such as Liu Chi,[c11] a learned technical general of the early Yüan[d] time. The bibliography and contents of this work, perhaps the most important of all for the history of gunpowder in Chinese culture, have been brilliantly elucidated of late by Ho Ping-Yü[e] and Wang Ching-Ning[f] in Australia. The various versions can be dated, I believe, between about + 1280, the end of the Sung,[g] and about + 1380, well after the establishment of the Ming. It thus covers the period of the Yüan Dynasty and the time when the new emperor to be, Chu Yüan-Chang,[h] was conducting his campaign to overthrow the Mongolian dominance, a campaign in which he made much use of guns and cannons, especially the new bombards. One of his gunners, Chiao Yü,[i12] was probably the ancestor of another of the same surname, Chiao Hsü[j13] who lived much later in the Ming, and I think both were associated with the *Huo Lung Ching* tradition.

Then you may turn to the *Wu Pei Chih*[k] (the Record of Arsenal Preparations), compiled by Mao Yüan-I[l] in + 1621, which is a very important work, also with many illustrations and also extant in several versions, sometimes with different titles. Besides all these primary sources, information about gunpowder weapons can also be found in other technical books—for example, the celebrated *T'ien Kung K'ai Wu*[m] (Exploitation of the Works of Nature) written by Sung Ying-Hsing[n] in + 1637; and, of course, there are always the encyclopaedias.

Now, the curious thing about this literature is that it looks both backward and forward. For example, there are insertions which are

[10] Chuko Liang, + 181−234, Captain-General of Shu, the renowned general and strategist of the San Kuo period.

[11] Liu Chi, + 1311−1375, technical general, helped Chu Yüan-Chang conquer the empire.

[12] Chiao Yü, ca. + 1345−1412, gunner and military writer, helped Chu Yüan-Chang conquer the empire.

[13] Chiao Hsü, late Ming and Ch'ing gunner and military writer.

[a] 火龍經 [b] 諸葛亮 [c] 劉基 [d] 元 [e] 何丙郁

[f] 王靜寧 [g] 宋 [h] 朱元璋 [i] 焦玉 [j] 焦勗

[k] 武備志 [l] 茅元儀 [m] 天工開物 [n] 宋應星

clearly anachronistic, such as pictures of bombards and culverins in the *Wu Ching Tsung Yao*, without any accompanying textual references; these pictures must have been put in by later editors. Conversely the *Huo Lung Ching* and the *Wu Pei Chih* illustrate and describe, presumably for the sake of completeness, a large number of gunpowder weapons which were almost certainly obsolete long before their time. Consequently in delineating the rise and development of gunpowder weapons we have to do a certain amount of conjectural reconstruction, arranging the different forms in the order most likely to have been that in which they actually appeared, aided now and then by certain dates which the texts themselves vouchsafe. This is the kind of reason which makes it difficult to say as yet with complete certainty that the final bombard stage appeared in China before it appeared in Europe; but it does really look as if the entire line of development, from the first mixing of sulphur, saltpetre, and a source of carbon to the metal-barrel gun and cannon, took place in China first and passed to Islam and Christendom only afterward. In any case, the principle of the gun barrel is unquestionably Chinese, and its origin lay in that natural tubing which has always been so convenient for all kinds of scientific and technical purposes, the bamboo stem.

Until now, nothing has been said in this lecture about the rocket. In this day and age, when men and vehicles have landed on the moon, and when the exploration of outer space by means of rocket-propelled craft is opening out before mankind, it is hardly necessary to expatiate upon what the Chinese started when they first made rockets fly. After all, it was only necessary to attach the bamboo tube of the fire lance to an arrow, in the reverse direction, and let it fly free, in order to obtain the rocket effect. Exactly when this great reversal happened has been a debatable question. Twenty years ago, when our contribution to *The Legacy of China* was written, we thought that rocket arrows were developed first about + 1000, in time for the *Wu Ching Tsung Yao*. Unfortunately the lack of an adequate descriptive terminology was deceptive in this matter, because this work gives drawings of *huo chien*,[a] "fire arrows," which look quite like later drawings of rockets, and these in their turn were also called *huo chien*. But as some of the *Wu Ching Tsung Yao* arrows are stated to have been launched like spears or javelins by means of an *atlatl* or spear-thrower, it is

[a] 火箭

unlikely that they were rockets, but rather tubes filled with incendiary substances designed for setting on fire the thatch and other roofs of the enemy city. This is not at all the first time we have encountered situations where a fundamentally new thing did not generate a new name. That was the case, for example, with hydromechanical clockwork. So the name *huo chien* stands for incendiary arrows and also for rockets, and this is confusing.

The discovery of the Tunhuang[a] banner of about +950 in one sense settled the question as to which came first, fire lance or rocket arrow. Although there are still some arguments for the first appearance of rocket arrows by +1000, it now seems that we may have to look in another direction for the beginnings of the rocket, and at a considerably later date. Toward the end of the +12th century, in the Southern Sung, there are descriptions of a firework used in some displays at court, the "earth rat," or *ti lao shu*,[b] a bamboo tube filled with a low-nitrate rocket composition and allowed to rush freely about on the floor. It was capable of frightening people, and we have a record that one of the Sung empresses was not amused thereby. This civilian employment would have reminded the wielders of fire lances of the recoil effect which they must always have had to withstand, whereupon someone tried a fire lance fitted backward on an arrow, with the result that it whizzed away in the air toward its target. That would have come about, we suppose, some time during the 13th century, for rockets were certainly well established as firearms during the Yüan time in the 14th.

Many further developments of great interest followed during the Ming and Ch'ing.[c] First of all, there were large two-stage rockets (surprising as it may seem, reminiscent of the Apollo spacecraft) in which propulsion rockets were ignited in two successive stages, releasing automatically toward the end of the trajectory a swarm of rocket-propelled arrows to harass the enemy's troop concentrations. Rockets were provided with wings and endowed with a birdlike shape, in early attempts to give some aerodynamic stability to the rocket flight. There were also multiple rocket-arrow launchers in which one fuse would ignite as many as fifty projectiles; later these were mounted on wheelbarrows so that whole batteries could be trundled into action positions like the regular artillery of a still later date.

[a] 敦煌   [b] 地老鼠   [c] 清

It is not generally known that rocket artillery played a considerable part in the military and naval history of the 18th and early 19th centuries in the Western world. The city of Copenhagen was set on fire by rockets from the British navy during the Napoleonic wars, and rocket troops were prominent in the days of the so-called Honourable East India Company, contending with princes like Tippoo Sahib. This was, however, a short-lived phase, for high-explosive shells and incendiary shells could be fired from more advanced artillery with much greater accuracy of aim, and the rocket batteries of the West died out after about 1850. Only in our own time did rocket propulsion come back into its own with the determination of man to leave the earth's atmosphere altogether. High explosives could not do anything to help that, in spite of Jules Verne's[14] vast cannon pointed upward at the moon.

In the National Military Museum in Peking there is a model of the two-stage proto-Apollo rocket with the host of small arrows to release at its destination. There is also a model of the bird rocket with wings.

As for the multiple rocket-launchers, the *Wu Pei Chih*[a] has many illustrations, from which one can see that they could fire thirty, forty, or fifty arrows at a time, which must have been quite alarming for troops on the other side. There is a model of one of these in the National Military Museum. Then there was the mounting of these rocket-launchers on wheelbarrows, four of them usually mounted on a single wheelbarrow, with a few extra spears thrown in, just as a kind of reminiscence. The rocket-battery soldiers also carried hand fire lances, presumably in case the other chaps got too near. There is also a model in the Military Museum, following the books, with two rows of launchers, and the fire lances also present. And then finally we have a whole row, a veritable battery, of these wheelbarrow rocket-launchers, half a dozen of them, all in a row.

The question then arises: What about transmission to the Western world? We can be very sure of one thing, namely that it must have occurred at some time during the second half of the 13th century. This was just the period of the massive penetration of

[14] Jules Verne, 1828–1905, French novelist and early science fiction author; wrote *Around the World in 80 Days, Journey to the Centre of the Earth*, and other stories.

[a]　武備志

Eastern Europe by the Mongolian people under Batu Khan,[a][15] yet parodoxically the Mongols seem not to have been responsible for the transmission. They valued gunpowder greatly later on, especially in the fighting which put Khubilai Khan[b][16] on the Chinese throne, but in their earlier phases, when as nomadic mounted archers and consummate horsemen they routed the knightly chivalry of Europe at the battle of Liegnitz[17] in 1241, firearms had not yet reached a high enough state of development to be useful for cavalry operations. Pistol, carbine, and revolver still lay far in the future. Thus, in my opinion, the probabilities lie in rather different directions.

Let us review for a moment the course of events in that turbulent 13th century. The Mongols were on the way up. First the Khwarizmian[18] lands were annexed, then the Jurchen Chin[c][19] Dynasty was overthrown in 1234, and far away to the west Mangu Khan[d][20] invaded Armenia in 1236. The following year saw the fall of Russian Ryazan,[21] and the Mongols invaded Poland. In 1241, along with the victory of Liegnitz, there was the siege and fall of Budapest, but also the death of Ogotai Khan,[e][22] to be succeeded by Mangu ten years later. Now around 1253 came the journeys of William de Rubruquis[23] and a number of other Franciscan friars to the Mongolian court at Karakorum.[f][24] They were diplomatic envoys more than missionaries, charged to seek the help of the Mongols against the Muslims, the traditional foes of the Frankish Christians. It was a classic case of that circling strategy by which one seeks

[15] Batu Khan died in + 1256. He led the Mongol expeditions against Russia, Poland, and Hungary in + 1237–1242.

[16] Khubilai Khan, + 1216–1294, became the "Great Khan" when his brother Mangu died in 1259, first Mongolian emperor of China.

[17] Liegnitz is forty-five miles northwest of Wroclaw (Breslau).

[18] Khwarizm, a state to the south of the Sea of Aral.

[19] The Jurchen Chin Dynasty, + 1115–1234, founded by the Nüchên, a Tungusic people whose traditional homelands were around the Amur River.

[20] Mangu Khan, brother of Khubilai Khan, died 1259.

[21] Ryazan, an area in central Russia, southeast of Moscow.

[22] Ogotai Khan, First Khakan from + 1229–1241; Chinghiz Khan's third son.

[23] William de Rubruquis, floruit + 1228–1293, a Franciscan friar, sent on a mission to the Mongol prince Sartak by Louis IX of France in 1253.

[24] Karakorum, site of the Mongol court in Mongolia.

a 拔都　　　b 忽必烈　　　c 金　　　d 蒙哥　　　e 窩闊台

f 喀喇崑崙

to mobilise the forces of allies or potential allies whose lands lie beyond those of one's immediate enemy. One would give a good deal to
know what exactly the Franciscans saw of gunpowder and firearms
during their wanderings in Mongolia and Cathay. Although such
interest consorted ill with their habit, they may have felt it their
duty to bring back knowledge and skills that might conserve the
safety and power of Christendom against the infidel. Thus, with
this transmission in mind, the activities of the friars need looking
at more closely than hitherto. One of them might even have been
accompanied by a Chinese gunner, who knew the multifarious
devices of the previous half-dozen centuries as well as the latest inventions, and was not averse to seeking his fortune in strange
foreign lands—but so far history has not heard of him.

As for the strategy, it succeeded beyond all expectation, apart
from the fact that the Mongols did it alone, did it for themselves,
and formed no alliances with the Christians. Having subdued Persia, they invaded Iraq beyond the Persian Gulf, and Baghdad fell
in 1258. Soon afterward the Mongolian Ilkhanate,[25] centred on
Iran, was established, and the great astronomical observatory of
Marāghah was founded. Then appears a second possible medium
of transmission, the travels of Rabban Bar Sauma[26] and his friend,
the account of which was translated from the Syriac long ago by
Wallis Budge.[27] These young men were two Chinese Christian
Nestorian priests, born and educated in Peking, who pined to go
on a pilgrimage to Jerusalem. Neither of them ever got there, but
they did travel the whole length of the Old World before one of
them returned home. The friend was unexpectedly elected bishop
and catholicos of all the Nestorian churches when in Tabriz or
somewhere in Persia, and his duties therefore detained him there
indefinitely; but Bar Sauma travelled on to the West, visited Italy
and was warmly received in Rome (where no inconvenient questions about doctrine were asked), and finally reached Bordeaux,
where he celebrated the liturgy in the presence of the King of
England. Eventually he got safely all the way back to China. The

[25] Mongolian Ilkhanate. The Arabs used the term Ilkhanate to express the
homage they owed to the Mongol Khan, a member of a foreign tribe (Turkish *Il*).
[26] Rabban Bar Sauma, Uighur Nestorian priest, born in Peking, who made
great travels in Europe in the 13th century.
[27] Sir E. A. Wallis Budge (1857–1934), Egyptologist, writer of many books on
Egypt and the ancient Near East.

purpose of this pilgrimage may also have been primarily political, or partly so, perhaps to get Western assistance for the Sung against the Mongols; if so, it never had the slightest chance of success, but once again our shadowy Chinese gunner might have come along with the two priests and handed on his knowledge to discreet persons in Europe who were capable of receiving it.

Lastly, in this same 13th century there were not only Franciscan friars and Nestorian priests but also, even more famous, the travelling merchants, of whom the most illustrious was of course Marco Polo, ''Il Milione,'' the man who affirmed that there were millions of ships on Chinese rivers and millions of bridges in Hangchou[a] — and fundamentally he was not far wrong. The crucial date at which he eventually left China was 1284. He has served Khubilai Khan for twenty years or so, sometimes on secret service missions, more often in the salt administration, and when he left it was by sea, accompanying a Chinese princess proceeding with a great fleet to wed some Middle Eastern potentate. This might have been an even more appropriate scenario for that Chinese gunner we have in mind, but unfortunately it is a little too late, for the gunpowder formula was first given to Europe at just about that same time by Roger Bacon (in an anagram) and Albertus Magnus,[28] a Franciscan and a Dominican respectively. However, Marco Polo was by no means the only Italian merchant in China during the 13th century. There was also Francesco Pegolotti,[29] who wrote a book on how to get there and back, a kind of Baedeker; and there was a whole settlement of European merchants and their wives at Yangchou,[b] to say nothing of the famous French artisan, Guillaume Boucher,[30] serving the Khan at Karakorum. So there are many possibilities and much may yet emerge therefrom. By 1355, the time when Chu Yüan-Chang was crowning his successes in China, the time is far too late, for the Europeans were certainly firing off bombards by 1327.

[28] Albertus Magnus, +1193−1280, learned Dominican friar who lectured on Aristotle in Cologne and Paris, one of the leading lights of mediaeval Western science.

[29] Francesco Pegolotti, agent for a great Florentine trading house, collected information about the trade routes from other travellers and published his book in +1340.

[30] Guillaume Boucher, floruit +1250, born in Paris, a goldsmith and mechanic at the court of the Great Khan.

[a] 杭州   [b] 揚州   [c] 朱元璋

The peak point at which we need to visualise our Chinese *huo shou*[a] as coming west lies rather between + 1260 and 1280, that is to say, a time at which both the eruptors and the true cannons in China were undergoing rapid development. I hope that further research will bring us more light. It may also be fruitful to consider the environment or accompanying circumstances in which the transmission occurred. From all our work we have been able to distinguish particular "transmission clusters," when several important inventions and discoveries came westward together. For example, there were several which accompanied the transmission of the magnetic compass and the axial rudder in the + 12th century, and there were others which went along with the blast furnace for cast iron, and the helicopter top, in the 14th. It remains to be seen what transmission exactly we should place with gunpowder in the 13th.

Next there is one more point which needs to be raised—cliché perhaps, an idée reçue, a vulgarism, or a false impression. The somewhat gloomy aspect of our whole subject is considerably relieved by the reflection that the oldest chemical explosive known to man has been of immeasurable importance, not only in war but also in the arts of peace. Without it the innumerable products of mining needed by modern civilisation could not have been won, without it the cuttings and tunnels that have been necessary for our lines of communication by river, canal, rail, and road could never have been formed. "What a pity it was," as Shakespeare wrote, "that villainous saltpetre should be digg'd, Out of the earth" to decimate the ranks of armoured knights and longbow men in Lincoln green; but Shakespeare was never able to converse with the engineers of the industrial revolution, who had a totally different idea about the value of explosives, and high explosives, too—the results of modern chemistry. We must therefore take a more balanced view of the discovery of explosives, and not be obsessed by their warlike, murderous uses. Now, the cliché to which I referred is one still often heard in the rest of the world— namely, that although the Chinese discovered gunpowder, they never used it for military weapons but only for fireworks. This is often said with a patronising undertone, suggesting that the Chinese were just simple-minded, yet it has an echo of admiration

[a] 火手

too, stemming from the chinoiserie period of the 18th century when European thinkers had the impression that China was ruled by a benevolent despotism of sages. And indeed it was quite true that the military in China were always, at least theoretically, kept subservient to the bureaucratic civil officials. Like scientists in England during the Second World War, they were supposed to be "on tap but not on top." So the cliché could have been right, but unfortunately it is not.

If we place the final experiments which led to the correct gunpowder formula (even low in nitrate) at some date between 800 and 850, then, as we know, the mixture was already being used as a slow match in the flame-thrower pump by 919, and it was fully operative in the rocket composition flame-thrower of 950. Of course it must have been used for recreational fireworks as well. So far as we are aware, no adequate history of fireworks in China has ever been written, though Amiot[31] did something in the 18th century, and Fêng Chia-Shêng[a][32] much more in our own. It is certain, however, that they flourished mightily at the courts of Sui[b] and T'ang,[c] with coloured lights and balls of flame, so we can say that the rocket composition gunpowder must have been employed in these displays as soon as it became available. We have already noticed, too, that it was especially during the Wu Tai[d] period that gunpowder came into its own as a military weapon. No sooner had the Sung Dynasty commenced, by about + 1000, than the semiexplosive gunpowder was being enclosed in bombs and launched through the air by trebuchets (or mangonels), those early forms of artillery based upon the swape and the sling. Of course there were grenades thrown by hand as well, but that did not mean that fireworks did not continue, and indeed China became preeminent for them, as Jesuits like J. J. M. Amiot found when they arrived there after 1584. So the two uses, civilian and military, went on side by side down to the present day.

Finally the question may be raised whether explosives were ever used preindustrially in traditional China. Here a difficulty arises because of terminology. In mining and engineering the practice of fire setting, that is, the splitting of rocks by heat to make them

[31] J.J.M. Amiot, Jesuit missionary interested in science and technology, was in China in 1774 when the Pope dissolved the Jesuit order.
[32] Fêng Chia-Shêng, eminent modern historian of technology.

^a 馮家昇   ^b 隋   ^c 唐   ^d 五代

more easily movable, is ancient. Thus when it is said, as for example in the *Ming Shu*ᵃ that a certain governor set *huo kung*ᵇ technicians to work clearing away rocky projections in order to make a river navigable, it may well be that gunpowder was used, though the technique may also have been only fire setting; so this is a question which needs more careful examination.

There are two final important points to be made about this Chinese development of the first chemical explosive known to man. First, it is not to be regarded as a purely technological achievement. Gunpowder was not the invention of artisans, farmers, or master masons; it arose from the systematic if obscure investigations of Taoist alchemists. I say "systematic" very advisedly, for although in the 6th and 8th centuries they had no theories of the modern type to work with, that does not mean that they had no theories at all. On the contrary, Ho Ping-Yüᶜ and I have shown that an elaborate doctrine of categories of affinity had grown up by the T'ang, reminiscent in some ways of the sympathies and antipathies of the Alexandrian protochemists but more developed, less animistic, and in fact looking forward to the days when chemists in Europe in the 18th century would be drawing up their famous tables of affinity.

Those first protochemists of Hellenistic times, whose writings are preserved in the *Corpus Alchemicorum Graecorum*, though very interested in counterfeiting gold and in all kinds of chemical and metallurgical transformations, were not as yet in pursuit of a "philosopher's stone" which would act as a medicine of immortality or an elixir of life. There is every reason to believe that the basic ideas of Chinese alchemy, which had been longevity-conscious from the beginning, made their way to the Latin West through the Arabic and Byzantine worlds. It is impossible, really, to speak of alchemy in the strict sense before the contribution of the Arabs, and, indeed, some people have even claimed that the word iself, *chemia* (*al-kīmīya'* in Arabic) and other alchemical terms are derived from Chinese linguistic roots.

Many pieces of chemical apparatus from the Han period have come down to us, such as bronze vessels with two reentrant arms, probably used for the sublimation of camphor. Certain forms of distilling apparatus, as I showed just now, are typically Chinese

ᵃ 明書   ᵇ 火工   ᶜ 何丙郁

and quite different from those in use in the West. One can easily imagine the Taoist alchemists mixing everything off the shelves, in all kinds of permutations and combinations, to see what would happen—whether perchance an elixir of life would be formed. Once saltpetre had been recognised and isolated, as it was at least since T'ao Hung-Ching's[a][33] time, about + 500, the inevitable was going to happen. To sum it up, the first compounding of an explosive mixture arose in the course of a systematic exploration of the chemical and pharmaceutical properties of a great variety of inorganic and organic substances, inspired by the hope of attaining longevity or material immortality. What the Taoists actually got was something else, but it was in its way also an immense benefit to humanity.

Second, and really last, in the gunpowder epic, we have another case of a socially devastating discovery which China could somehow take in her stride, but which had revolutionary effects in Europe. For decades, indeed for centuries, from Shakespeare's time onward, European historians have recognised in the first salvoes of the 14th century bombards the death-knell of the castle, and hence of Western military aristocratic feudalism. It would be tedious to enlarge upon this now. In one single year, 1449, the artillery train of the king of France, making a tour of the castles still held by the English in Normandy, battered them down one after another at the rate of five a month. Nor were the effects of gunpowder confined to the land. They had a profound influence also at sea, for in due time they gave the death-blow to the multi-oared, slave-manned galley of the Mediterranean, which was unable to provide gun platforms sufficiently stable for naval cannonades and broadsides.

Less well-known but meriting passing mention here is the fact that during the century before the appearance of gunpowder in Europe, that is, the 13th, its poliorcetic value had been foreshadowed by another less lasting development, that of the counterweighted trebuchet, also most dangerous for even the stoutest castle walls. This was an Arabic improvement of the projectile-throwing device, the *p'ao*,[b] or *huo p'ao*,[c] most characteristic of Chinese military art. It was not at all like the tor-

[33] T'ao Hung-Ching, + 451–536, famous Taoist alchemist and physician.
[a] 陶弘景   [b] 砲   [c] 火砲

sion or spring devices of Alexandrian or Roman catapults, but a simpler swape-like lever, bearing a sling at the end of its longer arm and operated by manned ropes attached to the end of the shorter one. This was the trebuchet mentioned earlier in connection with the first bombs in the Sung.

Socially, the contrast with China is particularly noteworthy. While gunpowder blew up Western military aristocratic feudalism, the basic structure of Chinese bureaucratic feudalism after five centuries or so of gunpowder weapons remained just about the same as it had been before the invention had taken place. The birth of chemical warfare had occurred, we may say, in the T'ang,[a] but it did not find wide military use before the Wu Tai[b] and the Sung,[c] and its real proving grounds were the wars between the Sung Empire, the Chin[d] Tartars and the Mongols in the 12th and 13th centuries. There are plenty of examples of its use by the forces of agrarian rebellions and it was employed at sea as well as on land, in siege warfare no less than in the field; but as there was no heavily armoured knightly cavalry in China, nor any aristocratic or manorial feudal castles, the new weapon simply supplemented those which had been in use before, and produced no perceptible effect upon the age-old civil and military bureaucratic apparatus, which each new foreign conqueror had to take over and use in his turn.

[a] 唐   [b] 五代   [c] 宋   [d] 金

3

Comparative Macrobiotics

Ever since Francis Bacon, historians of science have recognised that the origins of modern chemistry, as in the work of Robert Boyle,[1] Antoine Lavoisier,[2] Dalton,[3] Liebig[4] and so on, lay in ancient and mediaeval alchemy, where much of the apparatus and a great knowledge of the properties of substances were developed, though with different ends in view. People have generally thought that the first protochemistry was that which developed in Graeco-Egyptian Alexandria between the −2nd century and the +6th century. It was the work of people like Pseudo-Democritus,[5] Zosimus,[6] and Olympiodorus,[7] and it was transmitted to us in documents written in Greek. They are the contents of the famous *Corpus Alchemicorum Graecorum*, which comes in three volumes, and there are many of these fragments.

What is much less widely known is that there was a parallel Chinese tradition running at slightly earlier datings from the middle of the −4th century onward and accessible in extant Chinese texts. Tsou Yen,[a] for example, antedates Bolus of Mendes,[8] just as

[1] Robert Boyle, 1627−1691, one of the early Fellows of the Royal Society, father of chemistry, gave his name to one of the gas laws.

[2] Antoine Lavoisier, 1743−1794, French scientist, contributed much to early understanding of how oxygen took part in chemical reactions.

[3] John Dalton, 1766−1844, chemist, developed the atomic theory of chemical reactions.

[4] Justus von Liebig, 1803−1873, eminent chemist, one of the founders of organic and agricultural chemistry.

[5] Pseudo-Democritus, + 1st-century Hellenistic protochemist.

[6] Zosimus, + 4th-century Hellenistic protochemist and systematiser.

[7] Olympiodorus, + 6th-century Byzantine protochemist and writer.

[8] Bolus of Mendes, −2nd-century Hellenistic writer on natural history and protoscience.

[a] 鄒衍

Li Shao-Chün[a] antedates Pseudo-Democritus, while Ko Hung[b] and Zosimus, Sun Szŭ-Mo[c] and Stephanus of Alexandria[9] correspond closely enough. But the two traditions, in Hellenistic Alexandria and in the Ch'in[d] and Han[e] in China, were radically different, for whereas the Hellenistic one was devoted to "auriaction," the believed making of gold from other substances, the Chinese one was devoted to "macrobiotics," that is, the believed preparation of elixirs of immortal life; and for this the characteristically Chinese idea of material *hsien*[f] immortality was undoubtedly responsible —no other civilisation had it. The word "macrobiotics" is derived from Greek and comes from a phrase in Hippocrates where he says that life is short but the art is long, *ho bios brachys, hē techrē makrē*. *Macro* combines with *bios*, to make macrobiotics, which means the art of long life. A Latin-derived word could be "pro-longevity," the meaning of which is much the same, but macrobiotics is the word that we prefer.

In addition, in both civilisations there existed "aurifiction"— namely, the fraudulent making of imitation gold, gems, and other precious things from less valuable substances. Here again the Chinese had considerable priority, for the coining of false gold, *wei huang chin*,[g] was prohibited by an imperial edict of −144, while in the West the parallel is the rescript of Diocletian, one of the Roman emperors, in +293. These artisan traditions, by no means of course all necessarily fraudulent, were taken up and continued by the early mediaeval Latin West, but the Franks and the Latins knew nothing at all about medicinal chemistry until the time of Roger Bacon, toward the end of the +13th century.

There can be no doubt at all that the elixir concept reached Europe through the intermediation of the Arabic alchemists. *Al-iksir* itself is an Arabic word, "the medicine of man as well as metals," though Chinese etymologies for it have been more or less plausibly suggested. Then, when once the full macrobiotic elixir concept had added itself to that of the aurifactive "philosopher's stone," the way was clear in Europe for that great iatro-chemical movement of the +16th century led by Paracelsus, who in the view of many chemists and biochemists, including myself, deserves

⁹ Stephanus of Alexandria, +3rd-century Hellenistic protochemist.

a 李少君　　　b 葛洪　　　c 孫思邈　　　d 秦　　　e 漢　　　f 仙

g 僞黃金

to be placed, in spite of all his nonsense (and he had plenty of it), side by side with Galileo and Harvey, among the heralds of modern science. The most famous phrase that he coined was that "the business of alchemy is not to make gold, but to prepare medicines for human ills," and thus did the spirit of Li Shao-Chün and Ko Hung live on and flourish again right at the beginnings of modern science.

In the ancient world there undoubtedly existed a trans-Asian continuity, greatly enhanced after Alexander's conquests in -320, and further facilitated after Chang Ch'ien's[a] diplomatic and commercial expeditions in -110. Ostanes the Mede personifies it because the teacher of Pseudo-Democritus and those other Hellenistic chemists or protochemists was said to be a Persian, so that there you have an influence coming from Iran into the Hellenistic world from the East.

Probably we shall never be able to trace what the channels were that connected Tsou Yen with Bolus of Mendes and Pseudo-Democritus; we can only go on deepening our understanding of both of them. Probably the two foci of aurifiction and aurifaction, which centred primarily on Ch'ang-an[b] in the East and Alexandria in the West, were essentially independent. We do not know what influences there were between them. But some things undoubtedly came westward—probably, for example, the root of the word *chemeia*, that "goldery" which would have so much interested the merchants on the Old Silk Road. In Greek, *chemeia* has no satisfactory etymology, and the suggestion is that it derives from the Chinese word for gold, *chin*,[c] *kim*, *kem*, or *gum*, as it was pronounced in various dialects, which may be the origin of the root *chem*—that is why I talk about the "goldery." Another idea that came was that of the loves and hates of the elemental natures, a strengthening of the sexual Yin[d]-Yang[e] concept of chemical reaction as the birth of all novelty, and perhaps the idea of projection. What did not come at this time was the basic concept of the material immortality of the *hsien*,[f] the holy immortals. Nor did the belief in the elixir come, not until 1200 years later, because for that a different eschatology would have been needed, and then also the emphasis on time in Chinese alchemy had little echo in the West. The conviction of the value of mineral and metallic medicines did

---

[a] 張騫   [b] 長安   [c] 金   [d] 陰   [e] 陽   [f] 仙

Fig. 14. Traditional Chinese still type, from *Nung Hsüeh Tsuan Yao* (農學纂要).

not get through at that time, nor yet the scheme of natural forces in the *I Ching*,[a] which of course could not be exported until modern time. Similarly the death-and-resurrection motif of Greek protochemistry, the *prima materia* idea, never found its way to China. But it is possible that the idea of distillation did, though it could only have been a stimulus diffusion, since the design of Chinese distillation apparatus was so radically different. As for biological analogies, the West emphasised fermentation, whereas China emphasised generation.

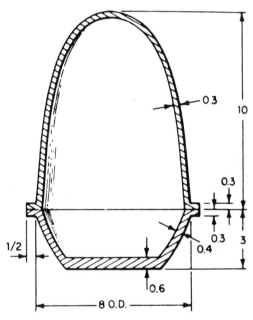

Fig. 15.   Conjectural reconstruction of Sun Szǔ-Mo's aludel or reaction vessel, ca. + 600.

However, there were common factors, like the majority of the chemical reagents they used (sulphur and mercury, and salts of various kinds) and the role of breaths or *ch'i*[b] like the *anathumiasis* of the Greeks, and other things of that sort in Nature's operations. Finally, a very interesting thing is that neither the Graeco-Egyptian protochemists nor the Chinese alchemists cared much about

<hr />

[a] 易經   [b] 氣

Fig. 16.   Distillation of mercury in retort, from *T'ien Kung K'ai Wu*, + 1637.

atoms. They left them to the Graeco-Roman philosophers, the Indians, and the Buddhists. It was all a pattern of imperfect communication; but the argument that there was no communication, and no will for it, cannot be sustained in the light of what we know now.

When between +635 and 660 the tribespeople of the Arabian deserts, inspired by the new religion of the Prophet Muhammad and determined to replace their poverty by a fuller life, poured forth into the surrounding areas of age-long culture, a fresh civilisation with its own language and its own characteristic features was born. It was destined, as we all know, to inherit the major part of Hellenistic science and technology and to pass it on in due course to the Latin West—a process of absorption, enrichment, and transference geographically facilitated by the fact that Islam conquered not only the Near and Middle East but also North Africa and Spain. Its cultural boundaries, to be sure, stretched much further to the East, reaching to the borders of India and the bounds of Sinkiang,[a] covering everywhere, in fact, as far East as the longtitude of Lop Nor[b] and all the space between Chad and the Caspian. Hence, it is easy to understand that Hellenistic knowledge was not at all the only river which flowed into the lake of Islam. Persian and Iranian tradition was swallowed up in it and strong currents of influence came westward, now from India and now from China. When Arabic began to concern itself with chemical matters, much was going to be added to the protochemistry of the Hellenistic world.

Now, Arabic alchemy does not really begin until the +9th century, but it may be significant that we have a circumstantial account of aurifaction seen by an Arab envoy at Byzantium toward the end of the previous century. His name was 'Umara ibn Hamza,[10] and being despatched on a mission by the Caliph al-Mansur[11] in +772, he was present at a demonstration in a secret laboratory in the imperial palace when lead was turned to silver by the projection of a white preparation, and copper to gold by the projection of a red one. The story is told by Ibn al-Faqih of Hamadan[12] about 902. He concludes that it was this incident

[10] 'Umara ibn Hamza, +7th-century Arab envoy interested in alchemy.
[11] Caliph al-Mansur, Abbasid ruler. Floruit +759–775.
[12] Ibn al-Faqih of Hamadan, floruit ca. +900, wrote prolifically.

[a] 新疆   [b] 羅布泊

Fig. 17.   Cupellation of argentiferous lead, from *T'ien Kung K'ai Wu*.

Fig. 18. Liquation process for the separation of silver from copper by lead, which is afterward cupelled, from *T'ien Kung K'ai Wu*.

which awakened the interests of the Caliphs in alchemy, and there
is no particular reason for disbelieving that, but whether aurifaction was really the first chemical exercise to intrigue the Arabs is
rather doubtful because the pursuit of macrobiotics was known at
least as early, and that must have come from a diametrically op-
posite quarter. It would have come from the East, from China, and
we know that people were talking about elixirs in places like Basra
already in + 690, quite an early time for the Arabic civilisation.
The great days of Arabic alchemy are reached with that flood of
books and tractates which go under the name of Jābir ibn Hayyān[13]
and can be dated with certainty to the last half of the + 9th century and the first half of the 10th. There has been a lot of trouble
about Jābir because, as you may possibly know, there was also a
Latin, ''Geber,'' who was supposed at one time to be a translation
of Jābir, but that is known now to be not the case at all. Geber
wrote in Latin toward the end of the + 13th century, about 1290,
and there is no connection between him and Jābir ibn Hayyān.
Geber's work shows no trace of having been translated from the
Arabic, and Jābir did not know many things which are in Geber, so
it is a totally different thing. The Jābirian writings form a corpus
almost like the *Tao Tsang*,[a] for there are about fourteen hundred
of them. They are the work of many different writers with a common philosophical outlook; none can be earlier than + 850, and
the whole collection must have been completed by about 930.
Whether there was any such person as Jābir ibn Hayyān himself is a
matter of controversy, but if he is accepted as historical, his dates
cannot have been far from 720 to 815, or perhaps some decades
later, and whether he himself wrote any of the texts in the corpus
remains also undecided.

   When you have a look at Arabic alchemy as it was in those times
you find yourself in a world very different from that of Hellenistic
protochemistry, even though Greek influences were manifold and
went very deep. Putting it epigrammatically, one could say that
aurifiction and aurifaction no longer dominate, for macrobiotics,
Chinese macrobiotics, and chemotherapy have come prominently
into the picture, together with biological products and substances,

---

[13] Jābir ibn Hayyān, floruit ca. + 776, has been called the father of Arabic
alchemy.

[a] 道藏

much more pharmacological interest, much closer connections with medicine, and a certain thread of preoccupation with all the phenomena of life. Theory also plays a greater role, and Arabic alchemy is therefore much more precise and logical than Hellenistic protochemistry, even though the structure is often based upon the most arbitrary assumptions, quite implausible to us today.

There is no need to go deeply into the question of Jābirian alchemy here, except to say that it adopted the four Aristotelian principles of heat, cold, moisture, and dryness, looking on them as real material constituents of things. Substances were believed to have external and internal characteristics—a substance such as gold, for example, was held to be hot and moist externally, but cold and dry inside. To convert one thing into another, as in this case silver to gold, it was only necessary to bring out the internal characteristics of the less noble metal. As for chemical change in general, everything depended upon the admixture or *krasis*, which the Arabs called *mizaj*, of the primary constituents, external and internal. The idea there was that as the balance of the constituents was perfect in gold, so if the balance of constituents, the *mizaj* or the *'adal*, could be made quite perfect in man, then he would be immortal, he would be one of the *hsien*, and he would not go down into the shades or into any of the other-worldly regions— which the Arabs, of course, believed in as People of the Book.

The agents for changing these balances and so converting one substance into something else by a transmutation were none other than the elixirs, *al-iksir*, among which one was supreme, and they were capable of neutralising constitutents present in excess and supplying the deficiencies of others.

At the same time, Jābirian Arabic alchemy was more advanced than Hellenistic protochemistry because of its clearer and more rational classifications. For example, they wrote of five spirits or volatile substances, seven metals, and a large number of pulverisable minerals, divided into vitriols, boraxes, salts, stones, and the like. Here the Arabs went beyond the Greeks, because to the classical sulphur, mercury, and arsenic a new volatile spirit was added—namely, ammonia, in the form of sal ammoniac. Indeed, Arabic writings are characterised from the start by a knowledge and use of ammonium chloride, *nushādir*, which came from natural sources in Central Asia, and of ammonium carbonate, *nushādir mustānbat,* obtained by the dry instillation of hair and other animal substances. There is no question whatever that they learned these things from China, because ammonium chloride occurs naturally in Sinkiang,[a] especially in volcanic regions, and it was the Chinese who told the Arabs about *nao sha.*[b] The names are so similar as to leave no doubt that *nushādir* was derived from *nao sha.*

A certain development of laboratory apparatus also occurred during these Arabic centuries, and another salt was added to those previously known—namely, potassium nitrate or saltpetre, *hsiao shih.*[c] That was the second most important substance which the Arabs learned about from the Chinese, and, as we saw, it was absolutely essential, being the basic limiting factor for the invention and use of gunpowder, the first chemical explosive known to man.

As for the volatile substances, the Arabic alchemists had the beginnings of the idea that all metals were combinations of sulphur, *al-kibrit,* and mercury, *al-zibaq,* in different proportions, or, if you like, *liu huang*[d] and *hung*[e] or *liu chu;*[f] all having naturally developed with great slowness in the bosom of the earth. Here again, though we cannot quote from a Chinese text any definite statement that all metals are composed of sulphur and mercury, nevertheless the overwhelming importance of these two substances in Chinese alchemy indicates very strongly that this was an idea which came out from China and deeply influenced the Arabic alchemists.

There are all sorts of interesting indications in the Arabic texts about things that came from China. For example, the Prophet Muhammad has a considerable number of *hadith* or undoubted sayings, and in one of these he said, "Go anywhere in quest of knowledge, even if it be as far away as China." And then al-Nadīm[14] in the section on alchemy in his *Fihrist* (+ 987) wrote the following words:

> I, Muhammad ibn Ishag, have lastly only to add that the books on this subject of alchemy are too numerous and extensive to be recorded in full, and besides the authors keep on repeating themselves. The Egyptians especially have many alchemical writers and scholars, and some say that that was the

---

[14] al-Nadīm, died + 995, wrote a "Catalogue of the Sciences" that contains much useful information about Arabic scientists and their books.

[a] 新疆   [b] 硇砂   [c] 硝石   [d] 硫黃   [e] 汞   [f] 硫珠

country where the science was born. The temples with their laboratories were there, and that was where Mary the Jewess worked. But others say that the discussions on the art originated among the first Persians, while according to others the Greeks were the first who dealt with it, and others yet again say that alchemy originated either in China or in India. "But Allah knoweth best what is the truth."

Thus by this time, in spite of the powerful influence of Hellenistic protochemistry, it could be thought quite possible that not only Iran but the cultures of East Asia could have produced the alchemical art. After that, it comes as less of a surprise to find that Hermes himself could be regarded as an inhabitant of China. Ibn Arfa' Ra's,[15] a Spanish Muslim of the +12th century, wrote an anonymous letter in which he said:

> The real name of Hermes was Ahnu [in other words, Enoch]. He was a dweller in the upper land of China, as the author of the "Particles of Gold" pointed out, where he said mining was looked after by Hermes in China, and Ares [probably Horus] found out how to protect the workings from flooding by water. Now, Ares lived in lower China and belonged to the first of the Indians. Furthermore he says that Ahnu, on whom be peace, came down from the upper to the lower land of China into India and went up a river valley in Serendib [that is, Ceylon], until he came to the mountain of that island where Adam, peace be upon him, descended. That was how he found the cavern, which he called the Cave of Treasures.

This anonymous letter is interesting because not only is it syncretistic in character, but it also mentions sal ammoniac, together with *diplōsis*, or "doubling" (the dilution of gold by alloying it with less noble metals, a favourite process of the Greeks). Then you find a little later, in a pre-Jābirian apocryphon, a discussion between a Byzantine emperor, Theodorus,[16] and an adept named Horus, who was said to come from lower China. Whether the expressions "upper China" and "lower China" have anything to do with Cathay and Manzi of the time of Marco Polo would be a subject on which much speculation might be permitted.

Another remarkable exchange occurs with regard to the invention of chemistry by a Chinese named Hua jen,[a] the changer. Rashid al-Dīn al-Hamdanī,[17] in his history of China completed in

[15] Ibn Arfa' Ra's, Arabic alchemist, +12th century.
[16] Theodorus, Byzantine emperor.
[17] Rashid al-Dīn al-Hamdanī, notable Arabic historian, died +1318.

[a] 化人

+ 1304, speaking of the time of the High King Mu[a] of the Chou,[b] mentions the exploits of the legendary charioteer, Tsao Fu,[c] and then goes on to say: "At that time lived a man called Hua jen. He invented the science of chemistry, and also understood the knowledge of poisons so well that he could change his appearance from one form to another in an instant of time." Now, there is no suggestion in these writings that Hua jen was anything but Chinese. In order to clarify Rashid al-Dīn's source you have to know two things: first, that he and his assistants were helped by two Chinese Buddhist physicians, Li Ta-Chih,[18] and someone whose Chinese name we do not know, and, second, that he depended on a little-known genre of Chinese historical writing—general surveys done from a Buddhist angle and incorporating the lives of Buddhas, *arhats*, *lohan*, and *bodhisattvas* within a framework of Confucian secular history. The first of these was the *Li Tai San Pao Chi*,[d] written by Fei Ch'ang-Fang[e19] in + 597, but the closest to Rashid's history was the work of a monk, named Nien-Ch'ang,[f20] entitled *Fo Tsu Li Tai T'ung Tsai*[g] (General Record of Buddhist and Secular History through the Ages), printed in + 1341.

In that book Nien-Ch'ang talks about Hua jen, and says that

> in King Mu's time a "Changer" appeared from the Far West. He could over-turn mountains and reverse the flow of rivers. He could remove towns and cities, pass through fire and water, and pierce metal and stone. There was no end to the myriad changes and transformations which he could effect and undergo. The king revered him as a sage and built for him a Tower of Middle Heaven to dwell in. Indeed, his appearance was like that of Manjuśri or Maudgalyayana or some such *bodhisattva*, but the king did not know that he had in fact been a direct disciple of the Buddha.

This story is familiar, because it is nothing else but a condensation and Buddhist adaptation of the opening part of the third chapter of the *Lieh Tzŭ*[h] book and therefore ascribable to any time between the −3rd and + 4th centuries. The Changer, Hua jen, was certainly not a historical person, but the chemical artisans did not appreciate such

[18] Li Ta-Chih, + 13th-century Chinese Buddhist physician.
[19] Fei Ch'ang-Fang, + 6th-century Chinese Buddhist historian.
[20] Nien-Ch'ang, + 14th-century Chinese Buddhist monk and historian.

[a] 穆王   [b] 周   [c] 造父   [d] 歷代三寶紀   [e] 費長房
[f] 念常   [g] 佛祖歷代通載   [h] 列子

fine distinctions, so it was natural that he should have become in due course the technic deity and patron saint of the art, craft, and science of chemical change.

Of course, when *Lieh Tzŭ* and other such books speak of the "Far West" they do not mean Europe or the Roman Empire, but rather that legendary land of the immortals, somewhere near Tibet or Sinkiang where reigned the great Queen Mother of the West, Hsi Wang Mu,[a] nothing short of a goddess. King Mu of Chou paid her a celebrated visit. That was the main theme, in fact, of the ancient book *Mu T'ien Tzŭ Chuan*,[b] and it was referred to in the *Lieh Tzŭ*. When centuries later the story came to the knowledge of real Westerners, like the group round Rashid al-Dīn, all this was omitted, and they took the Changer to have been a real Chinese with marvellous chemical knowledge. The significant fact that early in the + 14th century they were quite prepared to do this is the justification for telling you about him.

There are many things one can do to study Chinese-Arabic relations in this matter. One can note, for instance, the intensity of Arabic-Chinese intercourse over the centuries. One can observe how so many of the greatest scholars of the Islamic world came from countries on the borders of the Chinese culture area, and, though they made their fortunes in metropolitan Iraq or in Egypt, they may well have been recipients and transmitters of ideas current in their homelands.

Actually it was a surprise to us to find that for two or three centuries before the battle of the Talas River in + 741, the Chinese emperor had had a protectorate over all the lands of Khwarizm between the Iaxartes River and the Oxus right down to the Aral Sea. Many of these men, like the great algebraist al-Khwarizmī, came from that region. (One can always tell from their *nisba* where they came from.) It is a fascinating fact that such men would have been in contact with Chinese thought up there in Central Asia. Another thing one can do is to try and identify particular places on the Old Silk Road, which could have served as entrepôts for ideas. We cannot do all these things here, but I would like to relate a few stories about the *hu*[c] merchants, Persian and Arabic traders in China during the T'ang[d] and Sung.[e] Of course these merchants, whether themselves from Egypt, Iraq, Iran, or Central Asia, were

---

a 西王母   b 穆天子傳   c 胡   d 唐   e 宋

not confined to the foreign quarters in the great coastal ports, like Kuangchow,[a] for many of them came overland over the Old Silk Road, and since in T'ang times foreign people and things were all the rage, there was hardly a city in China unfamiliar with the *hu* merchants. *Hu* girls were also widely in service as dancers, maids, and entertainers, and there were *hu* grooms who can be seen in T'ang statuettes, attending to the horses and camels of the potentates of the T'ang. It was said of Ch'ang-an[b] that if one stayed there long enough one would meet representatives of every country in the known world. Not only Parthians, Medes, Elamites, and dwellers in Mesopotamia were there, but one could rub shoulders too with Koreans, Japanese, Vietnamese, Tibetans, Indians, Burmese, and Singhalese—and all had something to contribute on the nature of the world and the wonders thereof. If we want to know how they looked to the eyes of their hosts, we can find a good deal of information, because fortunately in +977, as part of a general programme of producing encyclopaedias and collections, the Sung emperor, T'ai-Tsung,[c] commissioned a treasury of rustic histories, biographical traditions, and short tales. This was in fact the *T'ai-P'ing Kuang Chi*,[d] edited by Li Fang[e] and completed in +978. Exactly how much is based on solid historical fact it is impossible to say now, and certainly a great deal of it was fiction, but in the present context that does not matter, for the text gives a clear indication of how the *hu* merchants seemed to the general run of literate Chinese scholars in the T'ang and Wu Tai.[f]

The *hu* were often concerned with alchemy and Taoism, skilled in recognising the gold of aurifiction or aurifaction, engaged in studying the art itself, and not only that, but concerned with life elixirs and physiological alchemy as well. So, as Edward Schafer[21] put it, the *hu* merchant in China was wealthy and generous, a befriender of young and indigent scholars, extremely learned in the knowledge of gems, minerals, and precious metals, a dealer in wonders, and not devoid of either magical or mysterious powers. So let us look at one or two individuals from among them.

In one report between +806 and 816, a young man named

---

[21] Edward Schafer, eminent American sinologist, wrote *The Golden Peaches of Samarkand* (1963) and other books on T'ang culture.

[a] 廣州   [b] 長安   [c] 太宗   [d] 太平廣記   [e] 李昉
[f] 五代

Wang Szŭ-Lang[a] masters the technique of making artificial gold, *hua chin*,[b] and saves his uncle from financial difficulties by giving him an ingot of it. Of this gold it is said that the Arabic and Persian merchants from the Western countries particularly wanted to buy it. It had no fixed price, and Wang used to ask what he liked for it. Another account, of +746, tells of a man called Tuan Yüeh[c][22] who met a merchant in a shop in Wei-chün. This merchant had more than ten catties of drugs, most valuable for the preservation of longevity and helping one to avoid cereal foods, *pi ku*.[d] Some of these drugs were difficult to get, however, and each day he used to go to the market to enquire of the Arabic and Persian merchants whether they had any. So here the *hu* apothecaries were directly involved in a trade related to the characteristically Chinese physiological alchemy. Then there was Li Kuan,[e] who, having received a beautiful pearl from a dying Persian merchant in recompense for his kindness, decided to place it in the mouth of the dead man, and many years later when the tomb was opened the body was found quite undecayed with the pearl still in position.

Another story was that of Lu,[f] and Li,[g] both Taoist adepts doing gymnastics and breathing experiences on T'ai-pai Shan.[h] One, having acquired great wealth by means of aurifaction, bestowed an alchemical staff on the other, which he said could be sold for a great sum at the Persian shop in Yangchow,[i] and so indeed it turned out. Evidently the *hu* merchants knew something valuable when they saw it.

And then there's the most remarkable story of Tu Tzŭ-Chün.[j][23] Tu, we are told, was an idle young scholar who met a strange old man in the Persian bazaar in the Western market in Ch'ang-an. Catching the old man's fancy, he was transferred from hunger and cold to a rich life of the utmost comfort. But before long it appeared that the stranger required his services for the accomplishment of an alchemical procedure designed to make an immortality elixir. There is a graphic description of Tu Tzŭ-Chün arriving at a remote place near Hua Shan,[k] fourteen *li* or so from the capital,

---

[22] Tuan Yüeh, +8th-century pharmacist.
[23] Tu Tzŭ-Chün, +8th-century alchemist.

[a] 王四郎	[b] 化金	[c] 段翌	[d] 辟穀	[e] 李灌	[f] 魯
[g] 李	[h] 太白山	[i] 揚州	[j] 杜子春	[k] 華山	

and there in the great hall was the old man attired in Taoist vestments, with an alchemical furnace nine feet high pouring out purple vapours, through which could dimly be seen nine jade maidens bearing the insignia of the cerulean dragon and the white tiger. But then the story takes a curious twist, for after ingesting certain drugs and sitting down to gaze in meditation at a blank wall, Tu found himself undergoing the torments of a variety of Buddhist hells, and was eventually reincarnated in another body before breaking the spell by a burst of uncontrollable emotion. Having failed to master these terrifying apparitions, Tu awoke, and the experiment which would have gained *hsien*[a] immortality for both him and the old Persian ended in failure.

Thus, taking all the evidence together, it seems clear that in the eyes of ordinary people, at least, during the T'ang[b] the merchants from Persia and the Arab countries were very interested in both the metallurgical and the macrobiotic aspects of Chinese alchemy. Looked at from this angle, it seems almost to stand to reason that Chinese ideas would have found their way westward to join with the Hellenistic ones that had been taken up into Arabic thought. The fact that nobody attempted the translation of an entire book of the *Tao Tsang*[c] into Arabic simply does not matter. One can imagine how fantastically difficult that would have been. All we are looking for is a new substance here and there, a few theories which might or might not have been understood, and one basic grand conception not misunderstood—namely, that chemical operations could perform miracles that were lifegiving and life-prolonging.

If you would like to make the personal acquaintance of a group of *hu*[d] merchant naturalists in China at this time, you could not do better than consider the Li[e] family in Szechuan.[f] Li Hsün[g][24] was of a Persian family which had settled in China during the Sui, and moved to Szechuan about +880. They were wholesalers, shipowners, and caravan patrons in the spice trade. Apart from his renown as a poet, Li Hsün became expert in materia medica, perfumes, and natural history, and wrote a book called the *Hai Yao Pên Ts'ao*[h] on the plant and animal drugs of the southern countries

[24] Li Hsün, +9th-century pharmacist.

a 仙   b 唐   c 道藏   d 胡   e 李   f 四川
g 李珣   h 海藥本草

beyond of the seas. This was similar to the + 8th-century *Hu Pên Ts'ao*[a] of Chêng Ch'ien,[b][25] but neither work has survived complete, I am sorry to say. Li Hsün's younger brother, Li Hsien,[c][26] was more of an alchemist, occupying himself with arsenical and other drugs, as well as essential oils and their distillation; he was also noted as a chess player. All this was in the time of the Ch'ien Shu[d] kingdom in Szechuan, and the younger sister of these two brothers, Li Shun-Hsien,[e][27] herself a poet of great elegance, became one of the leading ladies of that court. Probably unrelated to the family, though of the same name, was another *hu* physician, Li Mi-I,[f][28] who had sailed east to Japan in + 735 and had participated in the cultural Renaissance of the Nara[g] period.

Now comes a most extraordinary story. An important theme in Arabic alchemy which seems never to have been set properly in the context of elixir doctrine was the so-called "science of generation;" the *'ilm al-takwīn*.[29] This was concerned not only with the production of ores and minerals in Nature and the generation of the noble metals from the base, but also with the artificial asexual *in vitro* generation of plants, animals, and even human beings. You cannot just dismiss these ideas as mediaeval nonsense, because they often give deep insight into the minds of the men of that age, and illuminate what passed from one lot of men to another. Let us, therefore, take a look at this extraordinary development as it is found in its most explicit source, the *Kitāb al-Tajmī'*[30] in the Jābirian corpus. The artificial creation of minerals, plants, animals, and even of men and prophets by human artisan action, imitating the demiurge or creator of the world, was a cardinal belief in the Arabic 9th century. If you can succeed, said one writer, in composing or organising one particular thing, then you will assume the very place of the world soul in relation to substance. Isolated things will occupy in relation to yourself the place of the four qualities or

[25] Chêng Ch'ien, + 6th-century pharmacist.
[26] Li Hsien, + 9th-century alchemist.
[27] Li Shun-Hsien, court lady of the Former Shu Kingdom (Wu Tai).
[28] Li Mi-I, + 8th-century physician, went to Japan.
[29] *'Ilm al-takwīn*, the "science of artificial generation."
[30] *Kitāb al-Tajmī'*, the title means "The Book of the Concentration."

[a] 胡本草   [b] 鄭虔   [c] 李玹   [d] 前蜀   [e] 李舜絃

[f] 李密醫   [g] 奈良

natures, and thus you will be able to transform them into anything you wish. It sounds as if Hua jen[a] had come back.

Aurifaction was only one special case of this general principle. The possibility of an artificial generation of plants and animals was not confined to Jābirian circles but widely believed and discussed. So the whole matter has to be taken seriously, and the practical directions disclose some fascinating detail. For example, in one procedure a theromorphic glass vessel (that is to say, a glass vessel shaped like the animal you want to create) contained semen, blood, body parts of the organism to be reproduced, together with drugs and chemicals chosen in kind and quantity according to the method of the balance. All this was enclosed at the centre of a cosmic model, a celestial sphere, which the Arabs called *qura'* (and we should have called *hun i*[b])—globular, latticed, or armillary, set in continuous perpetual motion by a mechanical device. Meanwhile a fire of the first or unit intensity, that is to say a mild one, was kept burning underneath. If the exactly correct time was not reached, or if it was exceeded, no success whatever would be achieved. Obviously the exactly correct time never was reached, so that nothing ever happened—but that did not affect belief in the process. It was even averred that higher beings would come forth from the apparatus, equipped with the knowledge of all the sciences.

There can be no doubt that the origin of the famous *homunculus* of Paracelsus and Faust lies here, but how far Aldous Huxley would have been surprised to find his "Brave New World" of separated totipotent blastomeres and artificially incubated test-tube babies anticipated in the dreams of these Arabic alchemists we cannot undertake to say.

All these constructions are very un-Hellenistic, but they do signally recall the Chinese armillary spheres and celestial globes kept in continuous rotation by waterpower, instruments which belonged to the polar equatorial astronomy of China and not to the ecliptic planetary astronomy of the West. Similar Indian ideas especially concerning prepetual motion are also recalled. So much for the rotating cosmic shell.

As for the central vivification, the ingenuity of scholars has been much exercised to find Hellenistic antecedents, but there is not

[a] 化人   [b] 渾儀

much there except spontaneous generation, automata, and rituals for the animation of religious images, none of which is very much to the point. There were uncanny Graeco-Egyptian stories of speaking statues and ever-rotating columns, which of course came down to the Arabs, but here again honours are about even because Chinese culture also had a wealth of legends concerning automata, some of which, like the Taoist robot of King Mu[a] of the Chou[b] came very near indeed to being artificial flesh and blood. Once again there was not much to choose between Hellenistic and East Asian practices, for in China and Japan also there was the readying of images for the presence of gods, *lokapolas, bodhisattvas* and so on—even the insertion of model viscera to make them complete, as we find in some Japanese Buddhist statues still existing today. And then there was the formal consecration of images by the dotting in of the pupil of the eye. One can only conclude that the Arabs did not have to rely exclusively on Hellenistic culture for what they knew or thought they knew about spontaneous generation, mechanically operated simulacra, or the animation of religious images.

No, the fundamental feature of the Arabic creation of the rabbit out of the hat lay, as we see it, in those chemical substances which were added to the animal materials in the central container, for they represented nothing other than the *iksir* and *al-iksir* of life; and the entire pattern of pseudoscientific operations was simply a new and original Arabic exercise in the powers of the life-giving *tan*.[c] The Chinese elixir idea was at the centre, and the Chinese perpetual-motion cosmic model surrounded it. Beyond that, some parts were doubtless played by earlier Mediterranean ideas on the subject.

In general, therefore, this giving of life to the lifeless by chemical means was, we think, a particular Arabic application of a characteristically East Asian assumption: the giving of eternal life to the living by chemical means. It reminds me of Kungsun Ch'o[d] in the −4th century, saying with typical Chinese optimism: "I can heal hemiplegia, you know. If I were to give a double dose of the same medicine, I could probably raise the dead." Summing it all up, we think one could say that Arabic alchemical theory was a marriage between the Taoist idea of longevity or immortality,

[a] 穆   [b] 周   [c] 丹   [d] 公孫綽

brought about by the ingestion of chemical substances, and the Galenic rating of pharmaceutical potency, in accordance with the *krasis*, the *mizaj*, the *'adal*—the balance of the four primary qualities, the natures.

Arabic alchemists generally emphasised their ties with Hellenistic literature and traditions, and one does get that dominant impression in studying their writings; but perhaps if those were the books they read, the Persian and Indian and especially Chinese ideas and practices were what they really talked about, few or no texts from those lands being available in Arabic translation at any time. The macrobiotics of China seems to have come westward through a filter, as it were, leaving behind inevitably the concept of material immortality on earth or among the clouds and stars. After all, paradise for Muslims was quite similar to the heaven of the Christians. Nevertheless, some vital smaller molecules filtered through. There was, first, the conviction of the possibility of a chemically induced longevity, validated always by the example of the Old Testament patriarchs; second, the hope in a similar conservation or restoration of youth; third, speculation on what the achievement of a perfect balance of qualities might be able to accomplish; fourth, the enlargement of the life-extension idea to life donation in these artificial generation systems; and, fifth, the uninhibited application of elixir chemicals in the medical treatment of disease. Various writers have perceived that the whole course of Hellenistic protochemistry was primarily metallurgical—aurifictive and aurifactive, as we should say—whereas Arabic joined with Chinese alchemy in the profoundly medical nature of its preoccupations. Ko Hung,[a] T'ao Hung-Ching,[b] and Sun Szǔ-Mo[c] had glorious successors of the same cast of mind in al-Kindī,[31] the Jābirians, al-Razī,[32] and Ibn Sīnā.[33] If nothing living was really ever seen to step forth from Jābir ibn Hayyān's cosmic incubators, chemotherapy with all its marvellous achievements of today was certainly born from the Chinese-Arabic tradition with Philippus Aureolus Theophrastus Paracelsus Bombastus von Hohenheim as its great midwife.

[31] al-Kindī, Arabic alchemist (died ca. 873).
[32] al-Razī, famous Arabic alchemist and physician (+ 865–925).
[33] Ibn Sīnā (+ 980–1037), Avicenna, Arabic physician.

[a] 葛洪   [b] 陶弘景   [c] 孫思邈

Now, if the general picture I have been painting is about right it ought to be possible to find ideas of longevity, prolongevity, and macrobiotics in Byzantine regions. And, indeed, they are there, because in + 1063 Michael Psellus[34] wrote in his *Cosmographia* a very peculiar passage about the reign of the Empress Theodora, between + 1055 and 1056. Certain monks promised her an infinitely long life if she would follow their directions in various ways. Some of these monks were capable of walking in the air and doing all kinds of other things, like Taoist *hsien*, and they predicted for the empress a life going on for centuries without end, though in fact she died very soon afterward, in the second year of her reign, aged seventy-six. From this it seems quite clear that Theodora was under the influence of a group of Byzantine monks, who claimed to be in possession of macrobiotic techniques. The whole passage has a very Taoist, Sufi,[35] even Siddhi[36] character.

Some two hundred years later, Marco Polo was dictating to Rusticianus[37] his ideas about the *ciugi* (yogis) of India. I would like to quote a few words of his text. He says that

these Brahmins live more than any other people in the world [that is, they live longer], and that comes about through little eating and drinking, and great abstinence which they practise more than any other people. And they have among them regulars and orders of monks, according to their faith, who serve the churches where their idols are, and these are called yogis, and they certainly live longer than any others in the world, perhaps from 150 years to 200. Yet they are all quite capable in their bodies, so that they can go and come whenever they wish, and do all the service needed for their monastery and their idols, and though they are so old, they render it as well as if they were younger. And again I tell you that these yogis who live such a long time eat also what I shall explain, and it will seem a very great thing to you, a very strange thing, to hear. I tell you that they take quicksilver and sulphur, and mix them together with water and make a drink of them, and they drink it and say it increases their life and they live longer by it, and they do it twice in the week, and sometimes twice in the month. And you may know those people use this drink from their infancy so as to live longer, and without mistake those who live so long use this drink of sulphur and quicksilver.

---

[34] Michael Psellus, Byzantine historian interested in protochemistry.
[35] Sufi, the name of an ascetic and mystic sect of Islam.
[36] Siddhi, Tamil magicians.
[37] Rusticianus, a fellow-prisoner of Marco Polo who, in + 1298, wrote down Polo's description of his journey to China and residence there.

This passage is particularly interesting because the dietetic-hygienic element and the elixir-pharmaceutic element are both so prominently present. Li Shao-Chün's[a] cinnabar is living once again in Rusticianus's Latin. Now, Marco Polo was a contemporary of Roger Bacon. He reached China in 1275 and left for India in the year of Bacon's death, 1292, returning to Italy by 1295. Of course, Marco Polo's information did not spread with the rapidity of a mass-produced paperback at the present day, but it was widely read, and what he reported of the chemically induced longevity of Asian saints and sages must at least have chimed in with those other notes which emanated from specifically Arabic sources.

Last, I want to mention the first European who talked like a Taoist, and that was none other than Roger Bacon (1214–1292). Daringly did he affirm, time after time, that when men have unravelled all the secrets of alchemy there is no limit to the longevity they will be able to attain. This was but a part, of course, of his general scientific and technical optimism which makes him seem so modern a figure, so far ahead of his time. In addressing Pope Clement IV (who seems hardly to have taken very much interest), he says: "Another example can be given in the field of medicine and it concerns the prolongation of human life, for which the medical art has nothing to offer except regimens for healthy living. In fact there are possibilities for far greater extension of the span of life. In the beginning of the world the lives of men were longer than now, but life has been unduly shortened." In another place he writes: "The possibility of the prolongation of life is confirmed by the consideration that the soul is naturally immortal and not capable of dying, so, after the Fall, a man might live for a thousand years. Only since then has the length of life gradually shortened." So the shortening is accidental, and can be remedied wholly or in part. Of course he was referring to Methuselah,[38] our P'êng Tsu[b][39] in the West who lived for 969 years, but there is no doubt that he also took courage from other examples of the patriarchs. Then, in a paragraph full of burning enthusiasm, Roger Bacon ends as follows:

---

[38] Methuselah, mentioned in *Genesis* chapter 5 verse 27, is the Biblical type of a long-lived man.

[39] P'êng Tsu, the legendary long-lived man of Chinese tradition.

a 李少君   b 彭祖

And the experimental science of the future will know, from the "Secret of Secrets" of Aristotle, how to produce gold not only of 24° but of 30° or 40° or however many you desire. That was why Aristotle said to Alexander, "I wish to show you the greatest of secrets, and indeed it is the greatest; for not only will it conduce to the well-being of the States, and provide everything desirable that can be bought for abundant supplies of gold, but what is infinitely more important, it will give the prolongation of human life. For that medicine which would remove all the impurities and corruptions of baser metal, so that it should become silver and the purest gold, is considered by the wise to be able to remove the corruptions of the human body to such an extent that it will prolong life for many centuries. And this is that body composed with an equal temperament of the elements about which I spoke previously.

Here again are Ko Hung[a] and Jābir ibn Hayyān too in Latin dress, Franciscan dress, at least. The final sentence strikes at note familiar to us because Bacon explicitly reproduces the Arabic doctrine of perfect equilibration, immortality, and freedom from decay when all the elements are combined in perfect balance. We ought to quote, too, that passage at the end of Shakespeare's play *Julius Caesar*, when Mark Antony comes and finds the body of Brutus after the battle of Pharsalia: "His life was gentle, and the elements so mix'd in him, that Nature might stand up and say to all the world: 'This was a Man.'" In other books of Roger Bacon you get the same thing. For example, in the *Opus Tertium* of 1267 there is an interesting passage on speculative and operational alchemy, which treats explicitly of the generation of things from their elements, not only inanimate minerals and metals but also plants and animals. There was nothing very new in the belief that art could produce in a single day what Nature takes a thousand years to accomplish, but we must not miss the point that Roger was also extremely interested in the possibility of perpetual motion machines, probably to be achieved with magnets, like those his friend Pierre de Maricourt[40] was constantly occupied in attempting to make.

Finally, I shall end by saying a word or two about the real increase in longevity that has taken place through the centuries, since Roger Bacon's time. Although many other factors, such as food supplies, communications, housing, and sanitation have had leading parts to play in increasing the life expectancy at birth from

[40] Pierre de Maricourt, + 13th-century experimentalist in magnetism.

[a] 葛洪

twenty-four years for men and thirty-three for women in 1300 (Yüan time) to sixty-five for men and seventy-two for women in 1950, the rise of chemistry has also been outstandingly important. Greater chemical knowledge has undoubtedly led to a lengthening of human life; this is surely true beyond dispute. Seen from Ko Hung's point of view, all hygiene and bacteriology, all pharmacy and nutritional science, would have been but extensions of the chemical knowledge needed for preparing the *tan*.[a] The only failing of the early pioneers was the idea that there was one single substance which would be the universal medicine of men as well as metals, yet the elixir conception from Tsou Yen[b] through Jābir to Roger Bacon was veritably a great creative dream. The kernel of truth in it was that the human body has a chemistry of its own, like all other compounded bodies whether inorganic or organic, and that if man could gain deep knowledge of that chemistry he would be able to prolong his life beyond belief. If *hsien*[c] immortality still eludes us, one begins to wonder whether it always will—but what unimaginable changes in human society centuries hence will have to come about to control such knowledge, if indeed we ever attain it!

[a] 丹   [b] 鄒衍   [c] 仙

4

The History and Rationale of Acupuncture and Moxibustion

Acupuncture and moxa, as everybody knows, are two of the most ancient and characteristic therapeutic techniques of Chinese medicine. Acupuncture can be defined broadly as the implantation of thin needles to different depths at a great variety of points on the surface of the human body—points gathered in connected arrays according to a highly systematised pattern with a complex and sophisticated, if still essentially mediaeval, physiological theory behind it. In ancient times the technique was called *pien shih*[a] or *ch'an shih*,[b] and now *chên chiu*.[c] The needles unquestionably stimulate deep-lying nerve endings, hence their evocation of some far-reaching results. The classical theory was based on conceptions, still of great interest today, of a continuous circulation of *ch'i*[d] and blood *hsüeh*[e] around the body. I should like to dwell upon that because it is of particular interest, I think, for a university audience.

Moxa consists of the burning of *Artemisia* tinder, *ai*,[f] either in the form of incense-like cones, whether or not laid directly on the skin, or a cigar-shaped stick held just above it. The points chosen for application are in general identical with those of the acupuncture system, and there are expressions like *ai jung chih*[g] for that, or *ai jung chiu*.[h] Depending upon the degree of heat there can be either a mild thermal stimulus, like a fomentation, or alternatively a powerful counter-irritant cautery. Very broadly speaking, the acupuncture technique was from ancient times onward thought most valuable in acute diseases, while the moxa was considered more appropriate in chronic ones, and even for prophylactic purposes too.

[a] 砭石   [b] 鑱石   [c] 鍼灸   [d] 氣   [e] 血   [f] 艾
[g] 艾絨灸   [h] 艾絨灸

Now, acupuncture as a method of therapy, including sedation and analgesia, involving the implantation of needles, first developed during the Chou[a] period, in the −1st millennium. Today the needles are very thin, much thinner than the familiar hypodermic needles. Implantation into the body was made at different places at precisely specified points according to a charted scheme, based on ancient physiological ideas. The theory and the practice were, one finds, well established already, well systematised, in fact, in the −2nd century, though much development followed. We ourselves have many times seen the way in which the implantation with the needles is done, attending acupuncture and moxa clinics in several Chinese cities and in Japan, too; and one can say that the technique remains in universal use in China and in all Chinese communities at the present day. It also permeated centuries ago into all the neighbouring countries of the culture area, and for three hundred years past has awakened interest, together with a certain amount of practice, throughout the Western world. I suppose the first knowledge of it which the Western world had was in the book of Willem ten Rhijne[1] written in about 1684, and from time to time it has also been practised fairly widely in the West.

Dr. Lu Gwei-Djen[b] and I have a book in press at the moment which we have entitled *Celestial Lancets: A History and Rationale of Acupuncture and Moxa*. It is really part of *Science and Civilisation in China*, volume 6, part 3, but it is coming out ahead of time as a monograph. It will be useful, I think, because there is at present no regular history of acupuncture and moxa in any Western language. There are plenty of practitioners' manuals but no proper history and no real attempt to get down to the physiological and biochemical basis of either practice in terms of modern science.

The classical view of the circulation of the *ch'i*[c] was that it passed through a network of channels round and round the body. This was the *ching lo*[d] system, an arrangement of what we call acu-tracts and acu-junctions, twelve main tracts, and eight auxillary tracts, the famous *ch'i ching pa mo*.[e] You can readily see diagrams of these on charts or on manikins to illustrate it. Each of these tracts con-

---

[1] Willem ten Rhijne, + 1647−1700, served as a physician with the Dutch East India Company. He wrote a treatise on Asiatic leprosy (1687) and studies of tropical flora. The first to introduce acupuncture to the West.

[a] 周   [b] 魯桂珍   [c] 氣   [d] 經絡   [e] 奇經八脉

tains anything from ten to fifty acu-points (as we call them), the *shu hsüeh*,[a] at which the needles are implanted or over which the moxa is burnt. But besides the *ching lo* system of acu-tracts there is also the *ching mo*[b] system, which we translate as the "tract-and-channel network system," involving not only the circulation of *ch'i* in the tracts but also the circulation of blood in the blood vessels. It is extremely interesting to know that the idea of circulation existed so clearly and so long before the demonstration of the blood circulation by Sir William Harvey[2] (of my own College in Cambridge, I may say), in his *De Motu Cordis et Sanguinis in Animalibus* of 1628—but I will come back to that point later.

We also use the terms cheirotelic and cheirogenic, podotelic and podogenic, according to whether the tract starts from the hand or starts from the foot. A cheirotelic tract is one going to the hand and a cheirogenic tract is one coming from the hand; a podotelic tract is one going to the foot, and a podogenic tract is one coming from the foot. Now, each of these tracts is connected with one of the viscera, and this in a particular order: the lung, the large intestine, the gall bladder, and the liver. This connection was a great discovery of mediaeval Chinese physiology, because it involved what we now call viscero-cutaneous reflexes. For example, it is well known that appendicitis can be diagnosed from pain on pressure at McBurney's point, which is on the front of the body. And in many other cases things which go wrong with the viscera manifest themselves in pain points or other effects on the surface of the body. These useful signs are all viscero-cutaneous reflexes. It was a great achievement to become aware of them so long ago.

The cardinal importance of the system of acupuncture in the history of Chinese medicine is not denied by anyone, but its actual value in objective terms remained until recently, and to some extent still does remain, the subject of great differences of opinion. For example, you can find in various countries in East Asia modern-trained physicians, both Chinese and Occidental, who are sceptical about its value; but on the whole in China they are rather few, and the vast majority of medical men there, according to our experience, both modern-trained and traditional, do believe in its

[2] William Harvey, + 1578–1657, eminent physician, discovered the circulation of the blood.

[a] 輸穴   [b] 經脈

capacity to cure, or at least to alleviate, many pathological conditions. Now, nobody will ever really know the effectiveness of acupuncture (or of the other special Chinese treatments) until an adequate number of case histories have been analysed according to the methods of modern medical statistics. But that may well take a very long time, maybe fifty years or more, so great are the difficulties of keeping medical records in a country of a thousand million people, where the ratio of highly qualified physicians to the general population is low, and the need for medical and surgical treatment of all kinds so great and so urgent.

We decided that we could not wait that long in our work, and therefore we were obliged to embark upon our historiography with a mild preference in one direction or another. First, on the matter of published statistics it would not be fair to say that the Chinese medical literature contains no quantitative data. It does, and so does some of the literature in the Western world as well. However, the whole subject took a rather dramatic turn during the past fifteen years on account of the successes achieved in China in the application of acupuncture for analgesia in major surgery. Here there is no long and tortuous medical history to be followed up, no periods of remission or acute relapse, no chronic conditions with uncertain responses, no psychosomatic guesswork. Either the patient feels intolerable pain from the surgical intervention, or he does not, and the effectiveness can be known within the hour, or even sooner. More than any other development, this acupuncture analgesia (or anaesthesia, as it has often been called with undeniable but infelicitous logic) has had the effect of obliging physicians and neurophysiologists in other parts of the world to take Chinese medicine seriously, almost for the first time.

As for our preference, it derives, one might say, from a kind of natural scepticism. Both Dr. Lu and I are biochemists and physiologists by training, and as scientists of the modern sort we deal a good deal in scepticism; but scepticism can work in more ways than one. What we find hard to believe is that a body of theory and practice like acupuncture could have been the sheet anchor of so many millions of people for so many centuries when they got ill, if it had not had some objective value; and it strains our credulity as physiologists and biochemists to believe that the effects were wholly subjective and psychological. One might almost think in terms of a calculus of credibility, pending that time, which to us seems likely to be far ahead, when all the mysteries of psychosomatic causation will have been resolved. It seems to us more difficult to suppose that a treatment practised among so great a number of human beings for so long a time had no basis in physiology and pathology than that it was of purely psychological value. Of course you have to compare it with the practices of phlebotomy, or of bleeding, on a massive scale in the West, and urinoscopy, too, both of which practices had rather little physiological basis on which to sustain their extraordinarily long-enduring popularity, but neither of these had the subtlety of the acupuncture system. Besides, bloodletting had some slight value in hypertension, and extremely abnormal urines could also tell their story, though neither contributed very much to modern practice; however, I understand phlebotomy is having a comeback at the present day especially in dealing with high-blood-pressure cases.

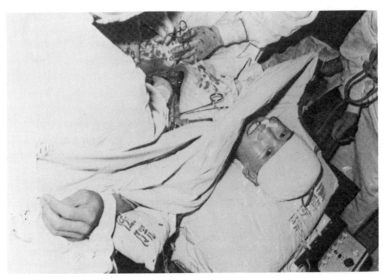

Fig. 19. Heart operation (commissurotomy for mitral stenosis) under acupuncture analgesia at the Shanghai Chest Hospital.

One view commonly expressed (mainly by Westerners) is that acupuncture has acted primarily by suggestion, like many other things in what they often call "fringe medicine," and some do not hesitate to equate surgical acupuncture analgesia with hypno-anaesthesia, in spite of many differences which we point out in our

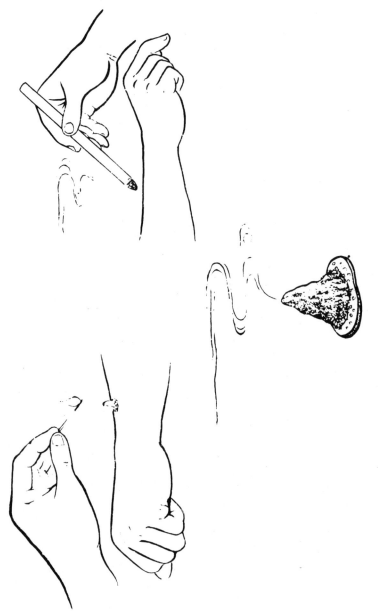

Fig. 20. Methods of moxibustion.

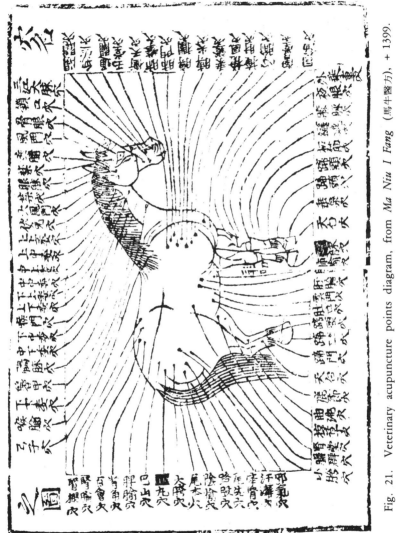

Fig. 21. Veterinary acupuncture points diagram, from *Ma Niu I Fang* (馬牛醫方), + 1399.

book. One outstanding difference which I may mention in passing has appeared only in the last year or so, and that is the effect of naloxone, which is a specific morphine antagonist. It is interesting to know that it does nothing whatever in hypnotism, and though it is certainly quite true that you can carry out major surgery under hypnotism, hypno-anaesthesia is not affected by naloxone. On the other hand, acupuncture analgesia is inhibited by naloxone; and therefore it almost certainly has something to do with the opioid peptides, the morphine analogues which we ourselves manufacture in our own brains. I shall come back to that again in a moment. Besides, to extend the term hypnosis to cover a general belief entertained by hundreds of millions of rational people for two millennia, and hence by prospective surgical patients today, would surely be a gross misuse of the term. Moreover, animal experiments, in which the psychological factor is largely ruled out, support our view that physiological and physico-chemical things are happening in the nervous system under acupuncture, and indeed animal experiments are increasingly being performed in laboratories investigating the technique. Not only this, but acupuncture has been a constituent part of veterinary medicine in China at least since the great treatises of the Yüan[a] period, the + 14th century, and continues to this day in widespread use for animal diseases.

Now, it is quite clear that in terms of neurophysiology the acupuncture needles stimulate various receptors at different depths, which send their afferent impulses up the spinal cord and into the brain. Perhaps they trigger events in the hypothalamus which activate the pituitary gland and lead to an increase of cortisone production by the suprarenal cortex, or perhaps they stimulate the autonomic nervous system in such a way as to lead to an increased output of antibodies from the reticulo-endothelial system. Both these effects would be of great importance from the therapeutic point of view, and in fact they are among the leading theories used at the present day to explain the therapeutic effects of acupuncture. In a way, it is easier to explain the analgesic effect than the therapeutic effect, but of course if cortisone production and the things allied with it are really stirred up by acupuncture or if antibody output is increased by it, it would be easy to see that acupuncture therapy could have been of enormous value, even in diseases such as typhoid and cholera where we know perfectly well the causative invading organism. On the other hand, in another situation, the needles may monopolise afferent input junctions in the thalamus, medulla, or spinal cord, in such a way as to prevent all pain impulses getting through to the cortex regions of the brain, thereby successfully inducing analgesia.

Space does not allow me to do more than mention the gating theories which have also been introduced to explain this kind of thing, but for the benefit of those who are not in the medical field, these theories can be introduced by a simple analogy: "gating" refers to a situation rather like that of a telephone exchange in which all lines are busy. If it can be arranged for all the lines to be busy all the time, then the engaged tone will ring the whole time, and this may be what is happening to the pain impulses in major surgery if acupuncture is ensuring that they do not get through to the cortex.

Furthermore, as I mentioned just now, there is no question that acupuncture analgesia is connected in some way, we do not yet know exactly how, with the opioid peptides of the brain. It was one of the most exciting and fascinating discoveries of the last five years, that our own brains manufacture substances known as enkephalins and endorphins, which can be no less than fifty times as powerful as morphine itself. Presumably the reason why they were not discovered earlier was that they are very rapidly destroyed by enzymes in the brain. But it was a rather interesting piece of research thinking, I believe, which pondered how it could be that Nature had invented both the morphine alkaloid and the morphine receptors in the brain, with the foreknowledge that man would one day discover the opium poppy. This could not really be believed, so scientists reasoned that if there is a morphine receptor somewhere in the brain, possibly in the thalamus or the reticular system, then what Nature really did was to make a receptor for something which the body itself produces. And this reasoning turned out to be absolutely right. Enkephalins and endorphins are very powerful analgesics, and they are made in our own brains, so here is another way in which you can see how the acupuncture needles at the periphery could stimulate neurons to fire off and release these highly powerful enkephalins or endorphins.

And then there are many other phenomena which ought to be

noted in physiology, for example the Head Zones. These have nothing to do with the head. They are named after Henry Head,[3] the great neurologist, because it was he who worked out the areas of innervation on the surface of the body which relate these various superficial regions to the viscera. This is connected with those viscero-cutaneous reflexes that I mentioned before, and it was he who showed in very ingenious ways the distribution of nerves on the surface of the body and how they connect with the viscera. I am glad that I had the honour of knowing Henry Head personally when I was a young man.

Another thing, too, is very relevant—the multifarious effects of "referred pain." I do not know how many of you experience this, but I have it constantly myself, in fact only this afternoon, as I was walking across the campus, I had a strong pain in my right Achilles tendon. I knew that it would go away in a minute or two because bubbles would come up in my intestinal tract, and as soon as the pressure on the wall of that was relieved, the pain in the foot instantly vanished. This is an example of referred pain. Some people experience it frequently, others do not have it so often, but this may well have been one of the phenomena which first stimulated the ancient Chinese physiologists to invent and draw out the system of acu-tracts.

Yet again some of the sensations experienced by patients undergoing acupuncture may have had an important effect on the formulations of the theory because, as you may know, there are four characteristic sensations experienced subjectively by patients undergoing acupuncture (*ma*,[a] *suan*,[b] *chang*,[c] and *chung*[d]). One of these (*ma*) involves what feels like a linear continuation of sensation. For example, if acupuncture is done at the *tsu san li*[e] point, just below the knee, the feeling of numbness will give the impression of travelling down to the foot, and the more the needle is twiddled by the acupuncturist, the more this will happen. So these lines of "radiation," as it were, from acupuncture points, may well have been one of the things which induced the idea of the acu-tract system in the first place.

Now, something remains to be said about the theoretical setting of acupuncture and other Chinese traditional methods, such as the

---

[3] Henry Head, eminent English neurologist.

[a] 麻   [b] 痠   [c] 脹   [d] 腫   [e] 足三里

medical gymnastics which are so important in China. Here I am talking about the relative value placed in Chinese and Western medicine respectively on aid to the healing and protective power of the body, on the one hand, as against direct attacks on invading influences, on the other. In both Western medicine and Chinese medicine these two concepts are to be found. In the West, besides the seemingly dominant idea of direct attack on the pathogen, we also have the idea of the *vis medicatrix naturae*, the natural healing power of the body. I am not likely to forget about that, because my father was still in general practice when I was small, and I remember how he used to talk to me about the *vis medicatrix naturae*. In those days, when I was about five or six, there were no antibiotics, no sulpha-drugs—there was diphtheria antitoxin, I agree—but in many cases the physician really had no recourse but to sit by the patient, and wait for the climax to be reached and the "crisis" to be overcome. So the natural healing power of the body was very important. That was a theme of resistance and the strengthening of resistance to disease which comes right down from the time of Hippocrates and Galen.

In China, where the holistic approach might have been thought to dominate, there was also the idea of combating external disease agents, whether these were malign or sinister *pneumata*, the *hsieh ch'i*,[a] from outside, of unknown nature, or distinct venoms or toxins left behind when insects had been crawling over food. That was a very old conception in China. So the attack on external agents was certainly present in Chinese medical thought too, and you could call it the *ch'u hsieh*,[b] getting rid of, or driving out, the *hsieh*[c] aspect; or, if you were a pharmacist in China, you would call it *chieh tu*,[d] dispelling the poison. The other one, the *vis medicatrix naturae* was surely largely what the Taoists meant in China by *yang shêng*,[e] the nourishment of life and the strengthening of it against disease.

It is fairly clear that whatever the acupuncture procedure does, it must be along the lines of strengthening the patient's resistance and not directly fighting invading organisms—that is, not the characteristic "antiseptic" technique which has naturally dominated in the West since the origin of modern bacteriology. In this connection, it is a significant fact that while Westerners are

[a] 邪氣   [b] 出邪   [c] 邪   [d] 解毒   [e] 養生

often prepared to grant value to acupuncture in things like sciatica or lumbago, for which modern Western medicine can do very little anyway, Chinese physicians have never been prepared to limit either acupuncture or moxa to such conditions; on the contrary they have recommended it and practised it in treating many diseases for which we all now believe we know the organisms (like typhoid and cholera), and they have claimed at least remissions if not radical cures. One of us, Lu Gwei-Djen, has vivid recollections of the dramatic recovery of her mother from cholera after acupuncture was done, and that was, I suppose, in Nanking in about 1910. The effect is thus, in principle, cortisone-like or immunological, and it is fascinating that both these ideas—the exhibition of hostile drugs and the strengthening of resistance—have developed in both civilisations and the medicine of both cultures.

In addition, there was a third idea, which sprang from the idea of balance or *krasis*. *Krasis* in Greek (*'adal* or *mizaj* in Arabic) was mentioned in the previous lecture. The idea was as much Chinese as it was Greek. It viewed disease as essentially a malfunction or imbalance, one or the other component entity in the body having unnaturally gained the lead over the others. Since the development of modern endocrinology, this conception has taken a new lease on life, but it was present from the beginning in both civilisations. European bloodletting and purging was a direct if crude result of it, because it was thought that "peccant humours" had to be got rid of; but in China a defective balance between Yin and Yang, or deviant relationships between the Five Elements, were generally diagnosed and altered in a more subtle way. Acupuncture was the first court of appeal here again. We have little doubt that many interventions of that kind did return the living human body with its nerves and hormones to a more even keel, though exactly how the mediaeval physicians visualised the interplay of the two great forces always remains rather difficult for us to understand in our time.

The universal reproach directed against therapeutic acupuncture by modern scientific medicine is the lack of statistical evidence. The absence of adequate clinical control experiments, the existence of the placebo effect, and the relative paucity of quantitative remission and follow-up data in contemporary China are hindrances, of course, but let no one say that the Chinese were unaware of the possibilities of spontaneous recoveries and remissions. There is a passage in the *Chou Li*[a] which talks about the official physicians of the imperial government. As you know, the *Chou Li* is an archaising treatise written in early Han[b] times about the kind of government bureaucratic organisation which ought to have existed in the Chou period, even if it never did. It was a kind of system of the ideal government organisation. Speaking of the I Shih,[c] the chief medical officer of the imperial government, it said: "The I Shih is in charge of the whole medical administration of the country, and he collects all efficacious drugs for the purpose of healing diseases. All those suffering from external maladies, whether of the head or body, are treated separately by appropriate specialists. At the end of the year he uses the records of each physician to decide on his rank and salary. Those who have cured 100 per cent of their patients are graded in the first class, those who have had 90 per cent recoveries are in the second class, those who have been 80 per cent successful are placed third, those who have cured 70 per cent are considered fourth class, and the lowest grade of all contains those who could not cure more than 60 per cent." Then the commentator in the +2nd century, Chêng K'ang-Ch'êng,[d][4] has a fascinating commentatory in which he says that "the reason why those who failed with four out of ten patients were placed in the lowest grade was because half of the cases might well have recovered anyway, even without any treatment at all." This text implies clearly the keeping of clinical records, and the comment seems to us an admirable example of the scepticism and critical mentality of the scholars of ancient China.

The following table is a distillation or evaporation of a very large amount of data which we had to take into account in the book that I spoke of before. I put there the statistical results of something like 150,000 acupuncture cases which one can collect from various sources, not only Chinese but also Russian and European. It is interesting, and perhaps surprising, that the relative success of both the therapy and the analgesia is about 75 per cent. The expression "cure" or "great relief" means in surgical terms that there is no need for any other kind of analgesic drug at all. "Marked relief" corresponds in surgery to its class II, and means that the acupuncture has to be supplemented either by pethidine beforehand or

---

[4] Chêng K'ang-Ch'êng, +2nd-century commentator on the *Chou Li*.

[a] 周禮   [b] 漢   [c] 醫師   [d] 鄭康成

some kind of sedative, or a chemical analgesic during the operation; this, therefore, is a second-class result. Then there are also
classes III and IV, denoting mild relief, no effect, and so on.
However, adding together classes I and II gives a figure of about
three quarters of the cases being successful.

Statistical results of acupuncture cases

I cure or great relief	II marked relief	I + II	III milder relief	IV no effect
therapy 44.1%	27.5%	71.6%	16.4%	12.0%
analgesia 37.3%	38.1%	75.4%	17.0%	7.6%
	placebo effect	30%-35%		

I particularly want to mention in connection with this table the
fact that the placebo effect exists, and always has to be taken into
consideration. It is an extraordinary fact, which not everybody outside medicine realises, that in an average population if you have a
person suffering from postoperative pain and you tell him or her
that you are going to inject a very effective analgesic, and that the
pain will go, and then you inject saline or something of that sort
with nothing effective in it at all, he or she may say that the pain is
greatly removed or has disappeared altogether. No less than 35 per
cent of the population will react like this—a phenomenon known
as the placebo effect. It is a significant thing for research because it
forms a kind of statistical baseline, which must be well surpassed
before it is possible to maintain that real results are at issue. In my
opinion, then, the demonstration in the above table that effectiveness of acupuncture in both therapy and analgesia stands at
about twice the placebo rate is a very important factor. One important discovery which has been made only in the last two years
is that the placebo effect itself is naloxone-sensitive. That must
mean, or at least it points to the conclusion, that the placebo effect
is in fact due to the mobilisation of the patient's own morphine
analogues. So it seems that given the psychological stimulus provided when the physician says "I am going to inject an analgesic,"
some 35 per cent of an ordinary population are able to respond by
mobilising their own enkephalins or endorphins.

At this point I would like to relate what I believe to be one of the most extraordinarily interesting stories in the whole of medical history: the story of circulation. It is, of course, well known that the *Huang Ti Nei Ching*[a]—*Su Wên*[b] and *Ling Shu*[c]—constitute the "Hippocratic corpus" of China. Not quite so old as the Hippocratic corpus in Greece, it is not much younger, and it consists of two parts: the *Su Wên* and the *Ling Shu*. We translate the former as "Questions and Answers about Living Matter," and the latter as "Vital Axis." In addition, there is another recension done in the + 7th century by Yang Shang-Shan,[d] called the *T'ai Su*.[e] Now, the *Su Wên*, which we date to the −2nd century, defines the blood vessels as the habitation of the *hsüeh*[f] (blood). From the time of the *Ling Shu*, onward in the −1st century, it was always said that the *Yin ch'i*[g] or *ying ch'i*[h] travels within the blood vessels (*mo*[i]), while the *Yang ch'i*[j] or *wei ch'i*[k] travels outside them. At the same time, the two *ch'i* were regarded as intimately connected. One of the commentaries on the *Nan Ching*[l] (The Manual of Difficult Cases) which is probably of the + 1st century, says that "the flow of the blood is maintained by the *ch'i*, and the motion of the *ch'i* depends on the blood. Thus coursing in mutual reliances, they move round and round." But this remark is not by any means among the more dramatic statements of the circulation which we have from these ancient times. For example, the *Ling Shu* says that what we call the vascular system, the *mo*, is "like dykes and retaining-walls, forming a circle of tunnels, which control the part that is traversed by the *ying ch'i*, so that it cannot escape or find anywhere to leak away." Commenting on this in + 1586, Wu Mao-Hsien[m] said: "It means that the *ying ch'i* travels within the blood vessels round and round, day and night, meeting nothing to stop or oppose it, and that is what the blood vessels in fact are." That is only the first of a number of instances which historically could not have been derived from William Harvey's famous publication of + 1628. But there is no need to quote from a Ming[n] text, because seventeen centuries earlier you can find statements like that in the *Su Wên*, where Ch'i Po[o] says that "the flow in the tract-and-channel network system runs on and on, and never stops,

---

[a] 黃帝內經   [b] 素問   [c] 靈樞   [d] 楊上善   [e] 太素   [f] 血

[g] 陰氣   [h] 營氣   [i] 脉   [j] 陽氣   [k] 衞氣

[l] 難經   [m] 吳懋先   [n] 明   [o] 岐伯

a ceaseless movement in an annular circuit [*huan chou pu hsiu*[a]]."
Clearly the circulation of the blood and *ch'i*[b] was standard doctrine
in the −2nd century, a situation contrasting amazingly with the
long uncertainty in the Western world, with its ideas (I should say,
"silly" ideas) of air in the arteries, and a tidal ebb and flow of the
blood.

The more detailed theory of the circulation can be found in the
Nan Ching, which says: "The *ying ch'i* runs within the blood
vessels, while the *wei ch'i* travels outside them in the acu-tracts.
The *ying ch'i* circulates endlessly, never coming to a stop except at
death. After fifty revolutions the two *ch'i* meet again, and that is
called the 'great meeting' [*ta hui*[c]]. The *Yin*[d] and *Yang*[e] *ch'i* go
along with each other in close relation, travelling in circular paths
which have no end [*ju huan wu tuan*[f]]. So one can see how the
ying[g] and *wei*[h] mutually follow one another." And then the com-
mentary goes on to explain that the fifty revolutions take place
during each day and night of twelve double-hours or one hundred
quarters. This the *Ling Shu* had already stated, pointing out that
the period corresponded both with the time taken by the sun to
traverse the round of twenty-eight lunar mansions (*hsiu*[i]), and
with 13,500 respiratory cycles of inspiration and expiration. Rough
measurements of the lengths of the tracts and principal blood
vessels gave a length of 162 feet as the distance to be traversed on
the complete cycle, so that the circulation repeated fifty times gave
a total distance of 8,100 feet (810 *chang*[j]), and the blood and *ch'i*[k]
must move forward just 6 inches during each individual respira-
tion. How tenaciously these figures were adhered to between the
−1st and the +16th centuries may be seen from a book such as the
Hsün Ching K'ao Hsüeh Pien[l] (Investigation of the 'Acu-points
along the Tracts), which reproduces them all precisely. Again,
since this was written in +1575, one can see that it was impossible
that the Harveian discoveries could have had any influence on this
classical Chinese tradition. I might mention also in passing that the
measurements on which they relied, the 162 feet and so on, came
from anatomical work done in the time of Wang Mang[m] between
the two Han dynasties.

[a] 環周不休	[b] 氣	[c] 大會	[d] 陰	[e] 陽	[f] 如環無端
[g] 營	[h] 衛	[i] 宿	[j] 丈	[k] 氣	[l] 循經考穴編
[m] 王莽					

What, then, was the position of the heart? It was summed up in the pregnant phrase *hsin chu mo*,[a] the heart controls the blood vessels. "The heart presides," says the *Su Wên*,[b] "over the circulation of the blood and juices, and the paths on which they travel." Wang Ping,[c] in the T'ang,[d] comments: "The heart controls the blood vessels, confining the *ying ch'i*[e] in its circular course; and the speed of its movement corresponds with the rate of breathing." Chang Chieh-Pin[f5] in +1618 (again *before* 1628), commenting further, said: "The heart rules over the circulation of the blood, and the pulse which it exhibits. In accordance with the motive power of the element Fire, it sends the blood to all the parts of the body." It seems likely, then, that through the centuries the heart must have been thought of as a pump of some kind, working in systole to propel the blood through its system of tubes. And we can find at least one clear analogy with the forge-bellows in pre-Harveian times, for in the *Lei Ching*[g] (The Medical Classics Classified) of Chang Chieh-Pin, mentioned just now, it is stated that "the heart and the pulse are not themselves either *ch'i* or blood, but rather the bellows of the *ch'i*[h] and the *hsüeh*.[i]" The statement "*Ch'i yü ch'i hsüeh chih t'o yo yeh*[j]" is really very interesting: *t'o yo* means bellows.

Since Chang Chieh-Pin was born in +1563 and brought out his compendium of medical physiology four years before Harvey's *De Motu Cordis*, it is evidently implausible to assume any influence of European ideas upon him, especially as the Harveian discovery had a long and uphill way to go before gaining general acceptance. Besides, Chang Chieh-Pin repeated the statement several times in his writings, so it may first have been made about 1593. Another pre-Harveian statement appears in the interesting words which Juan T'ai-Yüan[k6] used in the preface for the fourth edition of Matteo Ricci's world map of 1603. "I had the dim perception," he said, "that the earth is fixed and the air in motion, the water circulating with the air, somewhat as the blood and *ch'i* ceaselessly circulate, unresting, in the human body." Actually it is a little

---

[5] Chang Chieh-Pin, eminent Chinese physician. Born +1563, wrote *Lei Ching*.

[6] Juan T'ai-Yüan, Chinese scholar. Floruit +1600.

[a] 心主脈   [b] 素問   [c] 王冰   [d] 唐   [e] 營氣   [f] 張介賓

[g] 類經   [h] 氣   [i] 血   [j] 其猶氣血之橐籥也   [k] 阮泰元

unusual to find Chinese scientific thinking as late as this in advance of European, in contrast with earlier periods like the T'ang[a] and Sung,[b] when you come to expect it; but here there seems to be a particularly good case.

It is interesting to see how these ideas relate with those modern conceptions of blood circulation deriving from Harvey. On the classical Chinese estimate of fifty complete circulations in twenty-four hours, each revolution of the bloodstream would take 28.8 minutes. In the light of modern knowledge this was only about sixty times too slow, for the actual circulation time is approximately 30 seconds. However, Harvey himself never succeeded in getting that figure, which is a product of recent research. On the same page of the *De Motu Cordis*, where he reached the acme of his argument about the quantitative impossibility of the heart putting out so much blood in a given time, unless the blood got back to it somehow, through invisible channels, he added: "But let it be said that this does not take place in half an hour, but in an hour or even a whole day. In any case it is still clear that more blood passes through the heart in consequence of its action than can either be supplied by the whole of the food ingested, or than can be contained in the veins at the same moment." Harvey's emphasis was on quantitative reasoning backed by some measurements, just a few, and he remained very Aristotelian, giving a great deal of weight to analogies of microcosmic-macrocosmic type, but his great point was the quantitative one. He affirmed that the whole thing would not work unless the blood could get back to the heart somehow, to be pumped out again in due course. That was really his great contribution.

It is interesting to compare the data which would have been available to the *Nei Ching*[c] writers on the one hand and to William Harvey on the other. In ancient China they could easily relate the heartbeat time by the aid of delicate water-clock measurements, and by the same token they could give the respiratory frequency in terms of time. Moreover, as we have seen, they made measurements in Han[d] times of the approximate length of the great blood vessels, so they could form an estimate of the length of the course to be run. Lastly, they must have been familiar with the rhythmic spurting of blood from a severed artery, and since they

[a] 唐   [b] 宋   [c] 內經   [d] 漢

were predisposed on general philosophical and cosmological grounds to accept the idea of circulation, they must have assumed that somehow or other, in the intact body, the blood got back to the veins and the heart. Of course, they offered no experimental evidence to posterity, in the Renaissance manner; they simply stated their estimated conclusions about circulation time as part of general medical doctrine.

Harvey, for his part, was impressed basically by two things: first, the antiregurgitation valves in the veins, a piece of information which probably escaped the Chinese anatomists at all times, and, second, the quantitative measured amounts of blood sent forth by the heart which showed without question that it must get back there somehow. What is so incredible about the European experience is how long it took before the circulation was understood and accepted. And then there is the business about the heart as a pump, or a forge-bellows, as Chang Chieh-Pin[a] said. The earliest expression was one that struck me much, long ago when I was a young student at Caius, because we have William Harvey's own manuscript lecture notes there, and in his lectures he spoke of the heart working "as by two clacks of a water-bellows to raise water," but that remark is not contemporary with the lectures which he gave in 1616. It was a later insertion, probably made not before 1628. The next expression is in Harvey's *Anatomic Observations* of 1640, where he wrote that "the panting of the heart is but the pumping about of the blood, in the expansion receiving, and in the contraction sending it out." Various scholars have tried very hard to identify what kind of water pump Harvey had in mind. It seems likely to have been a contractile cylinder with corrugated leather sides, which was used widely as a fire engine in the 17th century. Nevertheless, it remains the case that pending further evidence the earliest reference to the heart as a pump is not European but Chinese.

The pump was, of course, a mechanistic conception, but it is generally agreed that Harvey is not explicable without his "dark side," that complex of cosmological ideas which partook of the nature of Hermetism, Neo-Platonism, and natural magic. He was a faithful Aristotelian, so he inherited the idea of the peculiar excellence of the circle as such which had so much inspired Giordano

[a] 張介賓

Bruno,[7] and he took his microcosmology seriously. For example, there were revolutions (that is, circulations) of the sun, moon, planets, and fixed stars in the heavens around some centre; there was the meteorological water cycle in the sublunary world; and there was the Prince, around whom everyone revolved in earthly States. Now if you compare Harvey's statements with those of the Chinese writers of Ming[a] and Ch'ing,[b] you see a considerable similarity. The chief difference is that the Chinese writers had behind them a steady tradition of blood and ch'i[c] circulation going back at least to the 2nd century. Walter Pagel[8] has studied the earliest European intimations of this, beginning with Plato himself, but the European statements are never as clear and explicit as those of the Chinese texts. Nevertheless, by the latter part of the Ming, toward the end of the + 16th century, there were leading intellectual figures in Europe who were asserting the circulation before Harvey's proofs. Giordano Bruno, for instance, in 1590 mentioned it in so many words, and in another writing of the same date he anticipated Harvey's parallel of the heart with the sun in the macrocosm.

Bruno is considered by many scholars to be a link of much importance between Andrea Cesalpino[9] and Harvey. Cesalpino in 1571 was the first of the anatomists to use the word *circulatio*, and it is well known that he described more or less correctly the pulmonary circulation. But here he had been preceded by several others, notably Raeldo Colombo[10] in 1559, and Michael Servetus[11] in 1546. Even more remarkable, he had been preceded by the Damascus physician Ibn al-Qarashī al-Nafīs,[12] who died in 1288. Ever since the first discovery of the relevant Arabic text there has been controversy as to whether or not this knowledge could have been transmitted to Harvey's 16th-century precursors. The weight of evidence that has now become available indicates that there was

[7] Giordano Bruno, floruit ca. + 1548–1599, Italian philosopher and theologian.
[8] Walter Pagel, eminent historian of medicine, wrote several books on Harvey, Paracelsus, and Renaissance medicine.
[9] Andrea Cesalpino, + 16th-century Italian anatomist.
[10] Raeldo Colombo, + 16th-century Italian anatomist.
[11] Michael Servetus, + 1511–1553, Spanish physician and theologian.
[12] Ibn al-Qarashī al-Nafīs, Arabic physiologist, d. + 1288.

[a] 明   [b] 清   [c] 氣

indeed a transmission, not only of the idea but of the arguments used to support it. One of the intermediaries has now been iden-tified as Andrea Alpago,[13] a Venetian consul-general, resident in the Levant for many years, and an accomplished oriental scholar who could read Arabic. What is more, al-Nafīs may have had more than an inkling of the general circulation itself, for he spoke of the aorta as the great vessel that circulates the animal spirits to all the organs of the body. Both the statement itself and its expression in terms of what we should call *ch'i*,[a] animal spirit, invite the ques-tion (or is it, I wonder, but a wild surmise?) as to whether even Ibn al-Nafīs and his contemporaries in the Arabic world could have been influenced by Chinese medical physiology. We cannot so far prove this; we have found nothing to suggest it in any of the pages of translations so far available, but it is established that Ibn Sīnā in the previous century was so influenced, especially in the matter of pulse lore. Much of what he said about sphygmology in the *Qanūn-fī al-Tibb*[14] came directly out of Wang Shu-Ho's[b][15] *Mo Ching*.[c] Moreover, in our studies of the spread of alchemical thought and practice we have found abundant evidence of transmissions from China to the Arabic and Western worlds. But we cannot yet say whether or not Ibn al-Nafīs was influenced by preceding Chinese thought, nor yet for sure whether those ideas were passed on by him to Cesalpino and Michael Servetus, and so reached the great Harvey.

When one looks back, one sees how great was the medical literature concerning acupuncture and moxa which grew up in China through the centuries. Many matters of absorbing interest arise as one looks through it. For example, first of all, leading on from what I have just been saying, there was the deep conviction of Chinese scholars and physicians about the circulation of *ch'i* and blood in the body, determining a rate of flow only sixty times slower than that which modern physiologists since Harvey have recognised, and that with a priority of about two thousand years. Then there was the discovery of the viscero-cutaneous reflexes, the connections of many parts of the body's surface with events occur-

[13] Andrea Alpago, + 16-century Venetian consul-general in the Levant.

[14] *Qanūn-fī al-Tibb* (Canon of Medicine), written by Ibn Sīnā.

[15] Wang Shu-Ho, + 265–317, Chinese sphygmologist, author of *Mo Ching* (Manual on the Pulse).

[a] 氣   [b] 王叔和   [c] 脈經

ring in the internal organs. Then, too, there was something I have not spoken of at all for lack of time, the appreciation of diurnal or circadian rhythms, and longer biological rhythms in man, and the development of an abstruse calculus on the basis of this to determine when acupuncture and moxa should best be performed. Finally there was the development of a very interesting module system for locating acu-points in human bodies of different sizes and proportions.

There have been a number of misunderstandings in the West about acupuncture and moxa. They have nothing whatsoever to do with parapsychology, occult influences, or psychic powers, and consequently do not deserve the praises of those who believe in such things. They do not depend entirely on suggestion, nor on hypnotic phenomena at all, and they are not contradictory of modern scientific medicine. Consequently they do not deserve the *odium theologicum* of the medical profession in the West. Acupuncture with moxa is simply a system of medical treatment which was already two thousand years old when modern science was born, and which has developed in a civilisation quite different from that of Europe. Today the explanations of its actions are being sought in terms of modern physiology and pathology. Great advances have been made already in this direction though the end is not yet in sight. It looks as if the physiology and biochemistry of the central and autonomic nervous systems will be the leading elements in our understanding, but many other systems, biochemical, neurochemical, endocrinological, and immunological are sure to be involved.

Another problem of great interest is the exact nature of the acu-points in terms of histology and biophysics. Since modern science did not spontaneously grow up in Chinese culture, acupuncture and moxa are traditionally based upon a theoretical system essentially mediaeval in character, though very sophisticated and subtle, indeed full of valuable insights and salutary lessons for modern scientific medicine. Again, the exact reinterpretation and reformulation of these theories, if such a thing is possible, will be a difficult matter for the future. However, we think it likely that in the oecumenical medicine of the coming years there will be a definite place for acupuncture both in therapy and in analgesia. Exactly how far this will be so, it is too early to say.

5

Attitudes toward Time and Change
as Compared with Europe

This afternoon I promised to talk about attitudes toward time and change in China and the West. I do not think I shall be able to tell you anything new, and I shall have to leave out a great deal since the time is too short for a general view of the whole situation regarding Eastern and Western attitudes toward time. (Incidentally, here I should say that nothing annoys me more than the use of the words "Eastern" and "Western," still worse "Oriental," because the Arabs, and the Indians and the Chinese, differed from each other really even more profoundly than the Europeans differed from some of them.) However, we must take a look at cyclical time and continuous time, and then I would like to say something about the deification of discoverers and about the recognition of ancient technological stages in time, one of the most important aspects of the history of science. Finally, we will discuss science and knowledge as cooperative enterprises cumulative in time; ideas of time in European civilisation, in Christendom, with regard to the coming of modern science; and Chinese ideas of time, too.

Now, the *philosophia parennis* of Chinese culture was, in our view, an organic naturalism which invariably accepted the reality and importance of time. This must be related to the fact that although metaphysical idealism is found in China's philosophical history, and even enjoyed from time to time a certain success (as when Buddhism was dominant in the Liu Ch'ao[a] and T'ang[b] periods, or among the followers of Wang Yang-Ming[c][1] in the + 16th century), it never really occupied more than a subsidiary place in Chinese thinking. Subjective conceptions of time were therefore uncharacteristic of Chinese thought.

[1] Wang Yang-Ming, + 1472−1528, official and China's greatest idealist philosopher.

[a] 六朝　　[b] 唐　　[c] 王陽明

Although of course we are speaking here of ancient and mediaeval or traditional thought, and not of sophisticated modern ideas, it may also be said that very clear adumbrations or relativism occur in the ancient Taoist thinkers. But whatever happened in time or times, whether flourishing or decaying, time itself remained inescapably real for the Chinese mind. This contrasts strongly with the general ethos of Indian civilisation (pointing up what I said just now), and tends to align China rather with the inhabitants of that other area of temperate climate at the Western end of the Old World.

Besides the Taoist, there were also, in the Warring States period, the Mohists and the logicians, the Mo Chia[a] and the Ming Chia,[b] who were remarkably advanced, as can be seen by comparison with the scientific thought of the Greeks. The Mohists were very near the formulation of the idea of functional dependence in relation to time and motion. Although the Stoics, with their great emphasis on the continuous ether rather than an atomistic universe, developed the first beginnings of multivalued logic and grasped one of the elements of the concept of function, they could not go much further, for they could not think of time as an independent variable with phenomena as its function. The description of motion by analytical geometry, as change of place functionally dependent on time, had to await the mathematisation of physics at the Renaissance.

For the Peripatetics, the followers of Aristotle, time was cyclical rather than linear, and in this they much resembled the Indians. They could never think of time as a coordinate stretching to infinity from an arbitrary zero—like the abstract coordinates of space, a geometrical dimension mathematically tractable—as Galileo[2] could and did. The Mohists had no deductive geometry like Euclid,[3] and certainly no Galilean physics, but their statements often give one a more modern impression than those of most of the Greeks. Why it was that their school did not develop in later Chinese society is one of the great questions which only a sociology

[2] Galileo Galilei, + 1564–1642, Italian astronomer and physicist who experimented with gravity and supported the Copernican theory that the earth revolves round the sun. The father of the Scientific Revolution.

[3] Euclid, 3rd-century Greek mathematician whose works, embodying deductive geometry, have formed part of the basis of Western mathematics.

[a] 墨家   [b] 名家

of science will be able to answer. For most of the followers of
Aristotle, however, there was something unreal about time, and
they were followed in this by most of the Neo-Platonists. In China
the Buddhist schools shared this conviction as part of their general
doctrine of the world as *maya* (illusion), but the indigenous
Chinese philosophers never did. I think that both Wang Yü[a] and
Liu Shu-Hsien[b] would agree with me in what I say here.

No classical literature in any civilisation paid more attention to
the recording and honouring of ancient inventors and innovators
than that of the Chinese, and no other culture, perhaps, went so
far in their veritable deification so late in historical times. Texts
which might be called technohistorical dictionaries, or records of
inventions and discoveries, form a distinct genre of literature.
Perhaps the oldest one of the kind is the *Shih Pên*,[c] most of which
simply recites the names and deeds of the legendary or semilegen-
dary culture-heroes and inventors, usually dubbed ministers of
Huang Ti,[d] systematising a body of legendary lore much more
copious than that of the technic deities of Mediterranean antiquity. Thus, Su Sha[e4] invented saltmaking, Hsi Chung[f] invented carts
and carriages, Chiu Yao[g] invented the plough, Kungshu Pan[h] the
rotary millstone, and Li Shou[i] computations. Various classes of
people are involved here (gods of antiquity demoted to heroes,
patronal deities of trades, mythical heroes euhemerised to inventors), then certain made-up names transparent in their etymology,
and lastly the true inventors who were undoubtedly historical per-
sonages, like the fourth of the examples I gave, also known as Lu
Pan,[j] who certainly did live in the Warring States period. The most
probable view is that the *Shih Pên* was first put together by
somebody in Chao[k] State between -234 and -228, just a little later
than the *Lü Shih Ch'un Ch'iu*.[l] From the post-Han centuries one
could find a dozen or more books to place in this category, and
writers were still not tired of it as late as the Ming when Lo Ch'i[m5]

---

[4] Su Sha, Hsi Chung, Chiu Yao, Kungshu Pan, and Li Shou, all legendary
culture-heroes.
[5] Lo Ch'i, + 15th-century archaeologist.

[a] 王煜　[b] 劉述先　[c] 世本　[d] 黃帝　[e] 宿沙　[f] 奚仲
[g] 咎繇　[h] 公輸般　[i] 隸首　[j] 魯班　[k] 趙　[l] 呂氏春秋
[m] 羅頎

wrote his *Wu Yüan*[a] (On the Origin of Things) sometime in the
+15th century.

So greatly prized was the love of traditional inventors that a list
of them was incorporated into one of the greatest arcana of Chinese
natural philosophy, the *I Ching*[b] (Book of Changes). This is a very
strange classic. It took its origin from what was probably a collection of peasant omen texts, accreted a large amount of material
concerned with ancient divination practices, and ended as an
elaborate system of symbols with their explanations. As we all
know, these were sixty-four patterns (*kua*[c]) of long and short lines,
in all possible permutations and combinations. Since to each of
these was assigned a particular abstract idea, the whole system
played the part of a repository of concepts for developing Chinese
science, the symbols being supposed to represent a gamut of forces
actually acting in the external world. The continuing additions to
the book made by many profound minds through the ages in the
form of appendices and commentaries turned it into one of the
most remarkable works in all world literature, and gave it enormous prestige in Chinese society, so that philosophical sinologists
are still studying it today with great interest. One, indeed, wrote
only a few years ago on the concept of time in the *I Ching*, showing
how inescapably this is bound up with its theme. Change is the
only thing in the universe which is unchanging. Nevertheless, others
have perhaps felt that on the whole the *I Ching* exerted an inhibitory effect upon the development of the natural sciences in
China, since it tempted men to rest in schematic explanations
which were not really explanations at all. It was in fact a vast, I
would almost say bureaucratic, filing system for natural novelty, a
convenient mental chaise longue which avoided the need for further observation and experiment.

The dating of the book is a very difficult question, but perhaps
the canonical text starts mainly in the −8th century, though it was
not completed until the −3rd, while the principal appended
writings, the "Ten Wings," must date from the Ch'in[d] and Han[e]
periods. One of these appendices makes a curious correlation between the great inventions and a select number of *kua*. Precisely
from these, it is alleged, the culture-heroes got their ideas. In other
words the scholars of the Ch'in and the Han found it necessary to

[a] 物原   [b] 易經   [c] 卦   [d] 秦   [e] 漢

educe reasons for the inventions from the corpus of the *kua* in the concept repository. Nets, textile-weaving, boat-building, houses, the craft of the archer, the miller, and the accountant—all are derived ingeniously from Adherence, Dispersion, Massiveness, Cleavage, the Lesser Top-Heaviness, the Breakthrough, and the like. What this teaches us here is chiefly, I think, the honour that was done to the venerated technic sages by incorporating them in the sublime world-system of the *I Ching*.

There was also much more concrete liturgical veneration. Everyone who spends time in China and travels about in the different provinces is deeply impressed by the many beautiful votive temples, dedicated not to Taoist gods or Buddhas or bodhisattvas, but to ordinary men and women who conferred benefits upon posterity. Some keep up the memory of great poets like the Tu Fu Ts'ao T'ang[a] in Ch'êngtu,[b] others that of great commanders like the Kuan Kung Lin[c] south of Loyang[d] but the technicians have a most eminent place. Twice in my lifetime I have had the privilege of burning incense, literally or metaphorically, to Li Ping,[e][6] that great hydraulic engineer and governor of Szechuan, of the -3rd century, in his temple of Kuan Hsien,[f] which stands and has for centuries stood beside the great cutting made under his leadership through the shoulder of a whole mountain. This venerable public work divided the main river into two parts, and still irrigates today an area fifty miles square supporting some five million people.

Every branch of science and technique is represented in these temples erected to the memory of doers and makers, deified by popular acclamation. The great physician and alchemist of the Sui[g] and T'ang,[h] Sun Szǔ-Mo,[i] had such a temple, and the custom continued even in the Ming, because Sung Li,[j] the engineer who made the summit levels of the Grand Canal a practical proposition, was given a votive temple posthumously beside its very water. Nor was incense burnt only to men. Huang Tao P'o[k] was a famous woman of the late +13th century, a textile technologist, instrumental in the propagation of cotton-growing, spinning, and weaving, which she brought to the Yangtze valley from Hainan. The towns and

---

[6] Li Ping, born in Ch'in State, greatest hydraulic engineer of Chinese antiquity.

[a] 杜甫草堂	[b] 成都	[c] 關公林	[d] 洛陽	[e] 李冰	[f] 灌縣
[g] 隋	[h] 唐	[i] 孫思邈	[j] 宋笠	[k] 黃道婆	

villages of the cotton areas all honoured her, and built many votive temples to her after her death.

It is thus impossible to maintain that the Chinese people had no recognition of technical progress. It may have proceeded at a leisurely pace, different from that which we have become accustomed to since the rise of modern science, but the principle is absolutely clear. I will return in a moment to the question of the recognition of the progress of human knowledge within the sciences themselves.

Meanwhile, we can see the idea of technical progress in another and very unexpected way. The conception of the three major technológical stages of man's culture, the ages of stone, bronze, and iron, following each other in a universal series, has been called the cornerstone of all modern archaeology and prehistory. In its modern form this was crystallised in 1836 by the Danish archaeologist, C.J. Thomsen, who used it to bring some order into the massive collections of the National Museum at Copenhagen, of which he was director. His good fortune was that during the subsequent decade the generalisation was placed for the first time on a fully scientific basis by the stratigraphical excavations of his compatriot, J.J.A. Worsae,[7] also in Denmark. It remains the basic classification of periods of high antiquity, and a permanent part of human knowledge.

For its general acceptance there were several limiting factors, but it had first to be acknowledged that stone-tool artifacts had indeed been made by man. It was necessary also to understand the correlation of orderly series of geological strata with time, and to escape from the prison of the traditional biblical chronology so as to recognise the archaeological evidence of man's true antiquity. Furthermore, it was necessary to link archaeological findings with some knowledge of the distribution of metallic ores and with some reconstruction of the most primitive techniques of copper, bronze, and iron production.

Nevertheless, Thomsen was only the nucleus of cystallisation, for the general idea had been in the air since the middle of the + 16th century, a time when curious enquirers into what they called *fossilia* were, as humanists, well acquainted with Greek and Latin texts. They certainly knew of the passage in part five of the *De*

[7] J.J.A. Worsae, 1821–1885, Danish archaeologist, wrote on the early history of Scandinavia.

Rerum Natura of Lucretius,[8] which distinguishes the three ages. It starts with the line *Arma antiqua manus ungues dentesque fuerunt*, and translates as follows:

> *Men's ancient arms were hands and nails and teeth,*
> *Stones too, and boughs broken from forest trees,*
> *And flame and fire as soon as known. Thereafter, force*
>   *of iron and bronze was discovered.*
> *But bronze was known and used ere iron,*
> *Since its nature is more amenable,*
> *And its abundance more . . .*

Lucretius was writing probably in the neighbourhood of –60, and those lines have been called just a general scheme of the development of civilisation based entirely on abstract speculation. But I am not so sure that Lucretius himself never picked up a flasked arrowhead! At any rate, his contemporaries in China were saying exactly the same thing and with no less appreciation of the rise of man in time from primitive savagery, with perhaps more sure and certain reasons for what they maintained.

The *Yüeh Chüeh Shu*[a] (Lost Records of the State of Yüeh[b9]) is attributed to Yüan K'ang,[c10] a scholar of the Later Han,[d] whose work, which certainly used ancient documents, was finished by + 52. Here, in a chapter on the swordsmiths, we find the following passage relating a discussion between the prince of Ch'u[e] and an adviser named Fang Hu Tzŭ.[f]

> The prince of Ch'u asked "How is it, I wonder, that iron swords can have the wonderful powers of the famous swords of old?" Fang Hu Tzŭ replied: "Well, every age has had its special ways of making things. In the time of Hsien Yüan,[g11] Shên Nung,[h12] and Ho Hsü[i], weapons were made of stone, and stone was used for cutting down trees and building houses, and it was buried with the dead. Such were the directions of the sages. Coming down

---

[8] Lucretius, about –99 to –55, Roman poet who in *De Rerum Natura* supported the philosophy and atomic theory stated by the Greek Epicurus.

[9] A feudal princedom absorbed by Ch'u in –334.

[10] Yüan K'ang, + 1st-century historical writer.

[11] Hsien Yüan = Huang Ti, the legendary Yellow Emperor.

[12] Shên Nung, the legendary folk-hero credited with teaching people the art of agriculture.

[a] 越絕書   [b] 越   [c] 袁康   [d] 後漢   [e] 楚   [f] 風胡子

[g] 軒轅   [h] 神農   [i] 赫胥

to the time of Huang Ti,[a] weapons were made of jade, and it was used also for other purposes such as digging the earth, and also buried with the dead. Such were the directions of the ancient sage kings. Then when Yü[b] the Great was digging dykes and managing the waters, weapons were made of bronze; with tools of bronze the I Ch'üeh[c] defile was cut open and the Lung-mên gate[d] pierced through. The Yangtze was led and the Yellow River guided, until they poured into the Eastern Sea. Thus, there was a communication everywhere and the whole empire was at peace. Bronze tools were also used for building houses and palaces. Surely all this was a most sagely accomplishment. Now in our own time iron is used for weapons, so that each of the three armies had to submit, and indeed throughout the world there was none who dared to withhold allegiance from the High King of the Chou.[e] How great is the power of iron arms! Thus you too, my prince, possess a sagely virtue!" The prince of Ch'u answered: "I see, thus it must have been."

So here, then, apart from the intercalation of a jade subperiod, possibly meaning stone of better quality, we have a sequence just as clear as that of Lucretius; and Yüan K'ang had two advantages. First he belonged to a distinct tradition. If we read the books of the Warring States philosophers we find time after time a lively appreciation of the stages which mankind had passed through in attaining the high civilisation of the late Chou. The Taoists and Legalists, from the −5th century onward, worked out a highly scientific version of ancient history and social evolution. They had at their disposal the ancient epics of Yao[f] and Shun,[g] enshrined in chronicles like the *Chu Shu Chi Nien*,[h] a text which came down from the Wei[i] State, just as the *Ch'un Ch'iu*[j] emanated from Lu;[k] and they had the lists of culture-heroes and inventors which ultimately formed the substance of the *Shih Pên*,[l] besides a great many mythological oral traditions. From these they made their culture-stage sequence with conscious reference to the customs of the primitive peoples around them. They spoke of men living in nests in trees (perhaps pile-dwellings), or holes in the ground, including caves; of the food-gathering stage and the origin of fire and cooked food; of the first making of clothes; of the development of the art of the potters; and of the first writings on bone and tortoiseshell. A passage in the *Han Fei Tzŭ*[m] book, relating a speech of Yu Yü[n] to the prince of Ch'in,[o] strongly suggests that the writer

[a] 黃帝   [b] 禹   [c] 伊闕   [d] 龍門   [e] 周   [f] 堯
[g] 舜   [h] 竹書紀年   [i] 魏   [j] 春秋   [k] 魯
[l] 世本   [m] 韓非子   [n] 由余   [o] 秦

had actually seen neolithic pottery, both red and black, as also the bronze vessels of the Shang,[a] cast in deep relief. Wood, stone, bronze, and iron were regularly associated, as in the passage we quoted just now from the *Yüeh Chüeh Shu*, and with one or another of the mythological rulers. One could easily write a whole book about what one might call this protoarchaeology.

Secondly, China differed from Europe in that the three technological stages had succeeded one another rather faster, and were thus almost parts of history rather than prehistory. Stone tools were still in general use in the Shang, and continued down to the middle of the Chou,[b] perhaps till the coming of iron, for it seems that bronze was little used for agricultural tools at any time. It is revealing that the physicians maintained a persistent tradition that in ancient times their acupuncture needles had been sharply pointed pieces of stone, the *pien shih*.[c] Neolithic cultures earlier than the Shang were known under the general name of a Hsia[d] period or kingdom, and it was well realised that they had had no bronze. Copper, tin, and bronze metallurgy, however, quickly reached great heights of expertise under the Shang, and the beautiful metal, the *mei chin*,[e] as it was called, remained in use for weapons and for marvellous sacrificial bronze vessels right down to the middle of the Chou. The introduction of iron then occurred in perfectly historical times, a little before the life of the Master, Confucius, toward the middle of the −6th century, and you can easily trace the profound social effects that it brought about.

Those, therefore, who have sought to dismiss Yüan K'ang's generalisation as cavalierly as that of Lucretius have had even less justification. "This is not a case," somebody wrote, "of genius forestalling science by two thousand years. An alert intelligence is simply juggling possibilities without any basis in fact or any attempt to test them." Actually, this is quite wrong, and neither of these alternatives is applicable at all. The scholars of the Chou and Han did not make stratigraphical excavation, it is true, but they had a far more secure basis for their conviction of the truth of the three technological stages than such a critic could conceive, for the very tempo of development of their civilisation had made them historians rather than prehistorians.

The time has now come to speak of the progressive development

_a 商　　_b 周　　_c 砭石　　_d 夏　　_e 美金

of knowledge, far beyond the level of ancient techniques. We are already thinking here about the recognition of a progressive development of human knowledge in time. It would be quite a mistake to imagine that Chinese culture never generated this conception, for you can find textual evidence in every period showing that in spite of their veneration for the sages, Chinese scholars and scientific men believed that there had been progress beyond the knowledge of their distant ancestors. The whole series of astronomical tables, about one hundred twenty of them from the middle Chou[a] to the Ch'ing,[b] indicates this point. They are usually called calendars (*li*[c]), but they are really ephemenies, like the *Nautical Almanac* published by the Royal Observatory at Greenwich, and therefore astronomical treatises in themselves. Unfortunately they have been unjustly neglected by Western historians of astronomy, including myself. Each new emperor wanted to have a new one, necessarily better and more accurate than any of those that had gone before. No mathematician or astronomer in any Chinese century would have dreamed of denying the continual progress and improvement in the sciences which they professed. One of our colleagues, a Japanese friend named Hashimoto Keizō,[13] is writing a book about the details of the increase in accuracy of the astronomical constants embodied by the Chinese astronomers in each one of those separate publications. The same may also be said of the pharmaceutical naturalists, whose descriptions of the kingdoms of Nature grew and grew and grew. You can see the number of main entries in the major pharmacopoeias between −200 and +1600, and when you plot them on a graph they show a remarkable growth of knowledge through the centuries. There was a big rise after +1100, probably because of increasing acquaintance with foreign (Arabic and Persian) minerals, plants, and animals.

The position in China would be well worth contrasting in detail with that in Europe. In his great work on the idea of progress, Bury[14] showed long ago that before the time of Francis Bacon[15] only

[13] Hashimoto Keizō, Japanese historian of science, writer on astronomical-calendrical subjects, on early machinery, and on Chêng Ho's ships.

[14] J. B. Bury, Professor of History at Cambridge, wrote *The Idea of Progress* in 1920.

[15] Francis Bacon, +1561−1626, philosopher and essayist.

[a] 周   [b] 清   [c] 曆

very scant rudiments of the conception of progress are to be found in Western scholarly literature. The birth of this conception was involved in the famous + 16th and 17th century controversy between the supporters of the "Ancients" and those of the "Moderns." The studies of the humanists had made it clear that there were many new things, such as gunpowder, printing, and the magnetic compass, which the ancient Western world had not possessed. The fact that these, and many other innovations, had come from China or other parts of Asia was long not known at all, but the history of science and technology as we know it was born at the same time out of the perplexity which this discovery had generated.

Bury had dealt with progress in relation to the history of culture in general. Edgar Zilsel,[16] much later, enlarged his method to deal with progress in relation to the "ideal of science." The ideal of scientific progress included, he believed, the following ideas: (1) that scientific knowledge is built up brick by brick through the contributions of generations of workers, (2) that the building is never completed, and (3) that the scientist's aim is a disinterested contribution to this building, either for its own sake or for the public benefit, and not for fame or personal knowledge or private personal advantage. Zilsel was able to show very clearly that expressions of these beliefs, whether in word or deed, were extremely unusual before the Renaissance, and even then they developed not among the scholars, who still sought individual personal glory, but among the higher artisanate, where cooperation sprang quite naturally from working conditions—people like Nicholas Tartaglia,[17] for example, the great gunner, or Robert Norman,[18] who made ships' compasses, and so on.

Since the social situation in the era of the rise of capitalism greatly favoured the activities of these men, their ideal was able to make some headway in the world. Zilsel traces the first appearance of the idea of the continuous advancement of craftsmanship and science to

[16] Edgar Zilsel, American historian of science and its methodological and philosophical implications.

[17] Nicholas Tartaglia (+ 1500–1557), Italian artillerist, discovered that the range of a gun is greatest when its deviation is 45°. He worked on many other aspects of military technology, and on mathematics.

[18] Robert Norman, English technologist, began life as a sailor, then turned to compass-making. His book on the lodestone, *The New Attractive*, appeared in + 1581.

Matthias Roriczer,[19] whose book on cathedral architecture appeared in + 1486. "Thus, science," says Zilsel, "both in theoretical and utilitarian interpretations, came to be regarded as the product of a cooperation for nonpersonal ends, a cooperation in which all scientists of the past, the present, and the future have a part." Today, he went on, this idea or ideal seems almost self-evident, yet no Brahamic, Buddhist, Muslim, or Latin scholastic, no Confucian scholar or Renaissance humanist, no philosopher or rhetor of classical antiquity ever achieved it. Zilsel would have done much better to leave out the reference to the Confucian scholars until Europe knew a little more about them, for in fact it would seem that the idea of cumulative, disinterested, cooperative enterprise in amassing scientific information was much more customary in mediaeval China than anywhere in the pre-Renaissance West.

Before turning to quotations in supporting that statement, we ought to recall that the pursuit of astronomy throughout the ages in China was not an affair of individual stargazing eccentrics; it was endowed by the State, and the astronomer himself was generally not a freelance but a member of the imperial bureaucracy with an observatory often located within the imperial palace. Doubtless this did harm as well as good, but at any rate the custom of cumulative teamwork was deeply rooted in Chinese science. Whole groups of excellent computers and instrument-makers gathered round the great figures, such as I-Hsing[a] in the + 8th century, Shen Kua[b] in the + 11th century, and Kuo Shou-Ching[c][20] in the + 13th. And what was true of astronomy was also true of naturalists, for many of the pharmacopoeias were commissioned by imperial decree; in fact, the first one so commissioned was the *Hsin Hsiu Pên Ts'ao*[d] of + 659, no less than one thousand years before the first royally commissioned pharmacopoeia in the West, the *Pharmacopoeia Londiniensis* of about + 1659. And we know of large groups who worked together at materia medica and the taxonomic sciences during their twenty-year compiling activities; for example, the team led by Su Ching[e] between + 620 and 660. In these respects the mediaeval scientists of China, building on the

[19] Matthias Roriczer, + 15th-century Austrian architect.
[20] Kuo Shou-Ching, + 1231−1316, eminent astronomer and official, expert in hydraulic and calendrical matters.

[a] 一行   [b] 沈括   [c] 郭守敬   [d] 新修本草   [e] 蘇敬

knowledge of their forbears, resembled quite closely the historians, who also came together in teams to produce some of the splendid large-scale works which we all know about.

Let me quote a few voices from the past to give colour to this, perhaps unexpected, attribute of Chinese science. Science is cumulative in that every generation builds on the knowledge of Nature acquired by previous generations, but always it looks outward to Nature to see what can be added by empirical observation and new experiments. "Books and experiments," wrote Edward Bernard[21] in 1671, "do very well together, but separately they betray an imperfection, for the illiterate is anticipated unwillingly by the labours of the ancients, and the man of authors deceived by story instead of science."

This theme of empiricism was very strong in the Chinese tradition. I love that passage in the *Shên Tzǔ*[a] book which says that "those who can manage the dykes and the rivers are the same in all the ages; they did not learn their business from Yü[b] the Great, they learned it from the waters." That was written probably in the +3rd century, and in the +8th you can find a book like the *Kuan Yin Tzǔ*,[c] where you read: "Those who are good at archery learned from the bow, and not from I[d] the Archer. Those who can think learned from themselves, and not from the sages." This is in part the message of that splendid story of Pien[e] the Wheelwright in *Chuang Tzǔ*,[f] who admonished his feudal lord, the prince of Ch'i, for sitting and reading old books instead of learning the art of government from personal knowledge of the nature of people, just as the artisan learns from personal knowledge of the nature of wood and metal.

Thus, always alongside the Confucian veneration of the sages and the Taoist threnodies about the lost age of primitive community there flourished these other convictions that true knowledge had grown, and would continue to grow, immeasurably more if men would look outward to things and build upon what other men had found reliable in their outward looking. *Ko wu chih chih*[g] —"The attainment of knowledge lies in the investigation of things." This was the pregnant phrase in the *Ta Hsüeh*,[h] a book

---

[21] Edward Bernard, +17th-century English writer.

^a 慎子   ^b 禹   ^c 關尹子   ^d 羿   ^e 扁   ^f 莊子

^g 格物致知   ^h 大學

later to become one of the classics, probably written by Yochêng K'o,[a22] a pupil of Mencius,[b] about −260. It became, as we all know, the watchword of Chinese naturalists and scientific thinkers all through the ages.

There is no Chinese century from which one could not cite quotations to illustrate the conception of science as a cumulative, disinterested, cooperative enterprise in time. K'ung Jung,[c23] who died in + 208, opined in a passage often quoted afterward that the ideas of intelligent men were often far better for their time than any of the sayings of ancient sages, and he illustrated his point by a reference to the application of water wheels to trip-hammer pounding-batteries for cereals and minerals. Already in about + 20, Huan T'an[d24] had traced the sequence of manpower, animal power, and waterpower in industry, a sequence hardly less significant than that of the three technological stages that we spent some time talking about just now.

In the field of astronomy and geophysics, Liu Cho[e25] appealed to the throne in + 604 for the authorisation of new research on solar shadow measurements, proposing the geodetic survey of a meridian arc. What he said was: "Thus, the heavens and the earth will not be able to conceal their form, and the celestial bodies will be obliged to yield up to us their measurements. We shall excel the glorious sages of old, and resolve our remaining doubts about the universe. We beg Your Majesty not to give credence to the worn-out theories of former times and not to use them." However, His Majesty did not agree, and Liu's wish was not granted until the following century, when that very remarkable meridian arc survey, twenty-five hundred kilometres long, was accomplished between + 723 and 726 under the superintendence of I-Hsing[f] and the Astronomer Royal at that time, Nankung Yüeh.[g] This did give results different from those previously accepted, and their descriptions show an enlightened recognition that the age-old beliefs

---

[22] Yochêng K'o, putative author of the *Ta Hsüeh*.

[23] K'ung Jung, + 153−208, an official, wrote *K'ung Pei Hai Chi*.

[24] Huan T'an, ca. −40 to + 25, author of *Hsin Lun*, criticised the emperor for believing in prophecies.

[25] Liu Cho, + 554−610, an official under the Sui and an early exploiter of interpolation formulae in astronomy.

a 樂正克   b 孟子   c 孔融   d 桓譚   e 劉焯   f 一行

g 南宮說

about the universe must necessarily bow to improved scientific observations, even though the scholars of former times, the *hsien ju*,[a] were discredited thereby. Again, at the end of the + 11th century, the idea of cumulative advance came up against the superstition that each new dynasty or reign period must make all things new. A new prime minister wanted to destroy the great astronomical clock tower of Su Sung.[b] There was doubtless an element of party politics here, but two scholar-officials, Ch'ao Mei-Shu[c][26] and Lin Tzǔ-Chung,[d] who warmly admired the clock and regarded it as a great advance on anything of the kind that had previously been made, exerted themselves to save it. This they succeeded in doing, and the great clock continued to tick on until the year of doom + 1126 when the Sung capital was taken by the Jurchen Chin[e] Tartars. The clock was transported away to their own capital near modern Peking, and reerected there. It still ran for a decade or two but before long the Chin Tartars had nobody able to repair it, so it stopped.

It is in connection with these astronomical clocks that we so often find the expression "nothing so remarkable had ever been seen before." This occurs, for example, in a description of a hydromechanical clock with elaborate jack work, constructed under the superintendence of the last Yüan emperor himself, Shun Ti,[f][27] Toghan Timur, in + 1354. It reveals the fact that Chinese scholars were very conscious of scientific and technical achievements, by no means always trivial in comparison with the works of the sages of old. It remains to be seen whether, when all the information is in, pre-Renaissance Europe was as conscious of the progressive development of knowledge and technique as they were.

In the light of all this, the widespread Western belief that traditional Chinese culture was static or stagnant turns out to be a typical Occidental misconception. It would, however, be fair to use the terms homoeostatic or cybernetic, for there was something in Chinese society which continually tended to restore it to its original character, that of bureaucratic feudalism, after all disturbances, whether caused by civil wars, foreign invasions, or inventions and

---

[26] Ch'ao Mei-Shu, + 1035−1095, scholar and official.
[27] Shun Ti = Toghan Timur, last emperor of the Yüan, came to power in 1333 and died in 1370.

[a] 先儒   [b] 蘇頌   [c] 晁美叔   [d] 林子中   [e] 金   [f] 順帝

discoveries. It is truly striking to see how earth-shaking were the effects of Chinese innovations upon the social systems of Europe when once they found their way there, yet they left Chinese society relatively unmoved. For example, in a previous lecture we talked about gunpowder, which in the West contributed so strongly to the overthrow of military aristocratic feudalism and the death of the feudal castle, yet after five centuries of use in China left the mandarinate (the civil service) essentially what it had been to start with. At the other extreme, the beginnings of Western feudalism had been associated with the invention of equestrian boot-stirrups, but in China, their original home, no such disturbance of the social order resulted. One may take the mastery of iron-casting, achieved in China some thirteen centuries before Europe obtained it, as another example. In China it was absorbed into customary usage for a great variety of purposes, both peaceful and warlike, but in Europe it furnished those cannons which destroyed the feudal castle walls I spoke of just now, and it formed the machines of the Industrial Revolution.

The simple fact is that scientific and technological progress in China went on at a slow and steady rate which was totally overtaken by the exponential rise in the West after the birth of modern science at the Renaissance. It has been said that it was during the Renaissance in the West, in the time of Galileo, that the most effective method of discovery was itself discovered, and I think this is probably a true way to put it. There was mathematisation of hypotheses about Nature, and a constant recourse to experiment to test them. What is important to realise is that although Chinese society was so self-regulating and stable, the idea of scientific and social progress and of real change in time was there. Hence, however great the forces of conservatism, there was no ideological barrier of this particular kind to the development of modern natural science and technology when the time was ripe, as it certainly is at the present day.

Lastly, we come to what is perhaps the greatest question of all that could be raised in this present context: namely, could there have been any connection between the differences, if any, in the conceptions of time and history characteristic of China and the West, and the fact that modern science and technology arose only in that latter civilisation? The argument set up by many philosophers and writers consists of two parts: first, the supposed demonstration that Christian culture was much more historically minded than any other, and, second, the view that this was ideologically favourable to the growth of the modern natural sciences during the Renaissance and the Scientific Revolution.

The first half of this argument has long been familiar ground for Occidental philosophers of history. Unlike some other great religions, Christianity was indissolubly tied to time, for the Incarnation, which gave meaning and pattern to the whole of history, occurred at a definite point in time. However, Christianity was rooted in Israel, a culture which, with its own great prophetic tradition, had always been one for which time was real and the medium of real change. The Hebrews were perhaps the first Westerners to give a value to time, the first to see a theophany, epiphany, in time's records of events. For Christian thought the whole of history was structured around the centre, a temporal midpoint, the historicity of the life of Christ, and it extended from the creation, through the *berith* or covenant of Abraham, to the *parousia*, or second coming of Christ, the Messianic millennium and the end of the world.

Primitive Christianity knew nothing at all about a timeless God, the eternal "is, was, and will be" (*aiōnōntōn aiōnōn*, "unto ages of ages," in the sonorous words of the Orthodox liturgies), its manifestation the continuous, linear, redemptive time process, the plan (*oikonomia*) of redemption. In this world outlook, the recurring present was always unique, unrepeatable, decisive, with an open future before it, which could and would be affected by the action of the individual, who might assist or hinder the irreversible, meaningful, directedness of the whole. A social purpose in history, the deification of man, the *theosis*, was thus affirmed; significance and value were incarnate in it, just as God himself had taken man's nature upon him, and died as a symbol of all sacrifice. The world process, to sum it up, was in other words a divine drama enacted on a single stage with no repeat performances.

It is customary to contrast this view sharply with that of the Greek and Roman world, especially the former, where cyclical conceptions dominated. In Hesiod we hear about successive ages repeating themselves, and their eternal recurrence is one of the few doctrines which it is quite certain that Pythagoras[28] taught. The

---

[28] Pythagoras, −6th-century Greek thinker, founded a school of philosophy, speculated about number, harmony, and cosmic order, and gave his name to a geometrical theorem.

other end of Hellenism saw the Stoic doctrine of four world periods and a fatalistic pietism of Marcus Aurelius.[29] Aristotle himself and Plato, too, were wont to speculate that every art and science had many times developed fully and then perished, so that time would yet again return to its beginning, and all things be restored to their original state. Such ideas were often combined, of course, with the long-term recurrences of observational and computational astronomy. Hence the notion, probably Babylonian, of the Great Year.

Now, cyclical recurrence precluded all real novelty, for the future was essentially closed and determined, the present not unique, and all time essentially past time: "That which has been is that which shall be, and that which has been done is that which shall be done, and there is no new thing under the sun." Salvation therefore could only be thought of as escape from the world of time, and this was partly perhaps what led to the Greek fascination with the timeless patterns of deductive geometry, to the formation of the theory of Platonic ideas, and to the "mystery religions."

Deliverance from the endless repetition of the wheel of existence at once recalls the world outlook of Buddhism and Hinduism; and indeed it does seem true that non-Christian Greek thought was extremely like that of India in this respect. A thousand *mahāyugas* (four thousand million years of human reckoning) constituted one single Brahma day, a single *kalpa*, dawning with recreation and evolution, ending with dissolution and reabsorption of the world spheres with all their creatures into the absolute.

The rise and fall of each *kalpa* brought ever-recurring mythological events. Victories of gods and Titans alternately, incarnations of Vishnu, churnings of the milky ocean to gain the medicine of immortality, and the epic deeds of the *Rāmāyana*[30] and the *Mahābhārata*.[31] Hence the innumerable incarnations of the Lord Buddha as told in the *Jātaka*[32] birth stories. The dimen-

---

[29] Marcus Aurelius, Stoic and Roman emperor who lived from + 121 to 180, wrote twelve books of *Meditations* in which he advocated tranquil acceptance of fate.

[30] *Rāmāyana*, one of the two great Hindu epic poems, originally composed perhaps in the −5th century.

[31] *Mahābhārata*, another great Indian epic.

[32] *Jātaka*, stories of the previous incarnation of the Buddha.

sion of the historical unique does not really seem to exist in Indian thought, with the result that it is generally agreed that India remained the least historically minded of the great civilisations; while in the Hellenic and Hellenistic situations, uninfluenced by Israel, only a few remarkable minds broke through the prevailing doctrine of recurrence—for example, Herodotus[33] the historian, and Thucydides,[34] and they only partially. Of course the hopelessness of this world outlook was greatly modified in India by the wisdom, more Hindu than Buddhist, of the duty of the householder and husbandman in his generation—in fact, its own kind of Stoicism, which gave to ordinary social life its honoured place in part at least of every individual's life cycle.

Paul Tillich,[35] that very great theologian, who settled in New York, brought together the characteristics of the two great types of world outlook into almost epigrammatic form. The Indo-Hellenic space predominates over time, for time is cyclical and eternal, so that the temporal world is much less real than the world of timeless forms, and indeed has no ultimate value. Being must be sought only through the fleshly curtain of becoming, and salvation can be gained only by the individual (of whom the self-saving *prateyeka buddha* is the prime example), not by the community. The world eras go down to destruction one after the other, and the most appropriate religion is therefore either polytheism, the deification of particular spaces, or pantheism, the deification of all space. It may seem this-worldly, concentrating on the passing present, but it dares not look into the future, and seeks lasting value only in the timeless. It is thus essentially pessimistic. For the Judaeo-Christian, on the other hand, time predominates over space, for its movement is directed and meaningful, witnessing an age-long battle between good and evil powers (here ancient Persia joined Israel and Christendom) in which, since the good will triumph, the temporal world is ontologically good. True being is immanent in becoming, and salvation is for the community in and through history. The world era is fixed upon a central point which gives meaning to the entire process, overcoming any self-destructive trend and creating something new which cannot be frustrated by

[33] Herodotus (−480 to −425), Greek historian and geographer.
[34] Thucydides (−460 to −400), Greek historian.
[35] Paul Tillich (1866−1965), eminent German-American Protestant theologian and philosopher.

cycles of time. Hence, the most appropriate religion is monotheism, with God as the controller of time and all that happens in it. It may seem other-worldly, despising the things of this life, but its faith is tied to the future as well as the past for the world itself is redeemable, not illusory, and the Kingdom of God will claim it. Thus, it is essentially optimistic.

I think we may surely accept this intense history-consciousness of Christendom. The second part of the argument, which appears to have been hinted at rather than worked out by philosophers of history, is that this consciousness directly contributed to the rise of modern science and technology during the Renaissance, and may therefore rank with other factors in helping to explain it. If it helps explain it in Europe, perhaps its absence elsewhere (or putative absence) might help explain the absence of the Scientific Revolution in those other cultures.

There can be no doubt that time is a basic parameter of all scientific thinking—half the natural universe, if only a quarter of the number of common-sense dimensions—and therefore that any habit of decrying time cannot be favourable to the natural sciences. Time must not be dismissed as illusory, nor depreciated in comparison with the transcendent and the eternal. It lies at the root of all natural knowledge, whether based on observations made at different times, because they involved the uniformity of Nature, or upon experiments, because they necessarily involve a lapse of time, which you may want to measure as accurately as you can.

The appreciation of causality, so basic to science, must surely have been favoured by a belief in the reality of time. It is not at first sight obvious, however, why this should have been more favoured by linear Judaeo-Christian rather than by cyclical Indo-Hellenic time, for if the time-cycles were long enough the experimenter would hardly be conscious of them; but still it may be that what the recurrence theories really sapped was the psychology of continuous, cumulative, never-completed natural knowledge, the ideal that sprang from the higher artisanate but came to fruition in the Royal Society and its virtuosi. For if the sum of human scientific effort were to be doomed beforehand to ineluctable dissolution, only to be reformed with endless toil, aeon after aeon, one might as well seek radical escape in religious meditation or Stoic detachment rather than wear oneself out day and night, engaged with one's colleagues in blindly constructing a reef on the rim of a submerged volcano.

Psychological strength was certainly not always weakened in this way, for otherwise Aristotle would never have laboured at his zoological studies. Nevertheless, it is probably reasonable to believe that in sociological terms, for the Scientific Revolution, where the cooperation of many men together (unlike the individualism of the Greeks) was part of the very essence, a prevalence of cyclical time would have been severely inhibitory, and linear time was the obvious background.

Sociologically the idea of linear time may have acted in another way also. It may well have strengthened the resolution of those who worked for a "root-and-branch reformation in Church and State," bringing into being thereby not only the "new, or experimental, science" but also the new order of capitalism. Must not the early reformers and merchants alike have believed in the possibility of revolutionary, decisive, and irreversible transformations of society? The concept of linear time could not, of course, have been one of the fundamental economic conditions to make this possible, but it may have been one of the psychological factors which helped the process. Change itself had divine authority, no less, for the new covenant had superseded the old, the prophecies had been fulfilled, and, with the ferment of the Reformation, backed by the traditions of all the Christian revolutionaries from the Donatists[36] to the Hussites,[37] people dreamt again apocalyptically of the foundation of the Kingdom of God on earth.

Cyclical time cannot contain apocalyptic time. In many ways the Scientific Revolution, however sober, however patronised by princes, had kinship with these visions. "That discouraging maxim *nildictum quod non dictum prius* [nothing has ever been said that was not said already before]," wrote Joseph Glanville[38] in 1661, "hath little room in my estimation. I cannot tie up my belief to the letter of Solomon; these last ages have shown us what antiquity never saw, no, not in a dream!" Perfection no longer lay in the

---

[36] The Donatists were members of a Christian sect of socialist character that arose in North Africa in + 311.

[37] The Hussites were followers of the Bohemian preacher and Christian socialist John Hus (+ 1373–1415), who was burned as a heretic at Constance.

[38] Joseph Glanville, + 1636–1680, English thinker who attacked scholastic philosophy.

past, books and old authors were laid aside, and instead of spinning cobwebs of ratiocination men turned to Nature with the new
technique of experiment and mathematised hypotheses, for the
method of discovery itself had been discovered.

As the centuries passed, the concept of linear time influenced
modern natural science more deeply still, for it was found that the
universe of the stars itself had had a history, and cosmic evolution
was explored as the background to biological and social evolution.
The Enlightenment also secularised Judaeo-Christian time in the
interests of the belief in progress which is still very much with us,
although today, when humanistic scholars and Marxists dispute
with theologians wearing coats of different colours, the coats (at
any rate, to an Indian spectator) are actually the same coats, worn
inside out.

This brings us lastly to consider the position of Chinese civilisation. Where did it stand in the contrast between the concept of
linear irreversible time and the myth of eternal recurrence? There
can be no doubt that it had elements of both conceptions, but
broadly speaking (and in spite of anything that can be said on the
other side) linearity, in my opinion, dominated. Of course, in
European culture the conceptualization of time was also an
amalgam, for although the Judaeo-Christian attitude was certainly
dominant, the Indo-Hellenic one never died out. One can see this
in the Spenglerian view of history in our own time, and it has
always been so. While Aurelius Augustinus (Saint Augustine)
worked out the Christian system of one-way time and history in
The City of God, Clement of Alexandria,[39] Minucius Felix,[40] and
Arnobius[41] were inclined to favour astral cycles like the *annus
magnus*, the Great Year. But none of that really amounted to
much in Europe, and I do not need to give other examples of the
same thing.

For China the case is similar. Cyclical time was certainly a prominent idea among the early Taoist speculative philosophers, in
later Taoist religion with its recurring judgment days, and in Neo-
Confucian thought with its cosmic, biological, and social evolution

---

[39] Clement of Alexandria, floruit ca. + 150–215, one of the Fathers of the
Church, notable for having applied Greek culture and philosophy in expounding
Christianity.
[40] Minucius Felix, + 2nd or 3rd-century African theologian.
[41] Arnobius, d. ca. + 330, African apologist.

ever renewed after the periodical ''nights'' of chaos. Later Taoist and Neo-Confucian concepts were undoubtedly influenced by Indian Buddhism, which brought to China the love of *mahāyugas*, *kalpas*, and *mahākalpas*.[42] But the early Taoist philosophy preceded this influence, and indeed we do not find in it any developed form of the doctrine. We find instead a poetic resignation based on acceptance of the cyclism of the seasons and the life spans of living things. But all this leaves out of account both the mass of the Chinese people throughout the ages, and also the Confucian scholars who staffed the bureaucracy, assisted the emperor in the rites of the age-old national Cosmism or Nature worship, and provided the personnel for the Bureaux of Astronomy and Historiography.

Sinologists have appreciated for more than a hundred years the linear time-consciousness of Chinese culture and its extraordinary success in the writing of history—greater perhaps than in any other culture. Thus, in an interesting paper, Derk Bodde[43] wrote:

> Connected with their intense preoccupation with human affairs is the Chinese feeling for time, the feeling that human affairs should be fitted into a temporal framework. The result has been the accumulation of a tremendous and unbroken body of historical literature, extending over more than three thousand years. This history has served a distinctly moral purpose, since by studying the past, one might learn how to conduct oneself in the present and the future ... This temporal-mindedness of the Chinese marks another sharp distinction between them and the Hindus.

What Bodde writes here about the great historical traditions of China, which envisaged love (*jen*[a]) and righteousness (*i*[b]) as incarnate in human history, and sought to preserve the records of their manifestation in human affairs, is certainly true. Its praise and blame (*pao pien*[c]) bias ''for aid in government,'' though somewhat of a limitation and liable to crystallise into dead convention, was poles apart from the idea of *karma* in the Buddhist faith. What it affirmed was that evil social results would follow evil social actions, and though these might lead to the personal ruin of an evil ruler, the effects might also, or only, be visited on his house or dynasty; but inescapable evil effects there would be. The system of

[42] *Mahākalpa*, a multiple of one thousand *kalpa* periods.
[43] Derk Bodde, Emeritus Professor of Chinese at the University of Pennsylvania, collaborator in SCC for social and intellectual factors.

[a] 仁   [b] 義   [c] 襃貶

rewards and penalties for good and evil actions, worked out through a series of reincarnations of a particular individual, was quite foreign. The Confucian historians were much more concerned with the community than the individual. If their time had not been linear, it is hardly conceivable that they would have worked with such historical-mindedness and such bee-like industry. Moreover, theories of social evolution, technological ages initiated by inventive culture-heroes, and appreciations of the cumulative growth of human science, pure and applied, are in no way missing from Chinese culture.

It would easily be possible, finally, to overestimate comparatively the Judaeo-Christian idea of the keying of time's flow to a particular point in space-time when an event of world significance occurred. The first unification of the empire by Ch'in Shih Huang Ti^a in -221 was a never-to-be-forgotten focal point in Chinese historical thinking, all the more important because of the unity of secular and sacred, which no political dichotomy between pope and emperor ever broke up. If you want something still more numinous, the life of the sage, the Teacher of Ten Thousand Generations, Confucius, the supreme ethical moulder of Chinese civilisation, the uncrowned emperor, whose influence is vitally alive still today, was at least as historical as that of any great ethical and religions teacher of the West or the Middle East. That the Confucian outlook was essentially backward-looking is a thesis which in the light of the evidence I have tried to bring forward this evening cannot be sustained. The Sage's Tao^b was not put into practice in his own generation, but his assurance was that men and women would live in peace and harmony whenever and wherever it was practised. When this faith, less other-worldly than Christianity (for $T'ien\ Tao,^c$ Heaven's Way, was not, strictly speaking, supernatural) joined with the revolutionary ideas implicit in Taoist primitivism, the radically apocalyptic dreams of $Ta\ T'ung^d$ (the Great Togetherness) and $T'ai\ P'ing^e$ (the Great Peace and Equality), dreams that men could and did fight for, began to exert their potent influences. Paul Tillich wrote: "The present is a consequence of the past but not at all an anticipation of the future. In Chinese literature there are fine records of the past but no expectation of the future." Once again, it would have been much better

^{a} 秦始皇帝 ^{b} 道 ^{c} 天道 ^{d} 大同 ^{e} 太平

not to come to conclusions about Chinese culture while Europeans still knew so little about it. The apocalyptic, almost the messianic, often the evolutionary, and (in its own way) the progressive, certainly the temporally linear—these elements were always there, spontaneously and independently developing since the time of the Shang[a] kingdom, and in spite of all that the Chinese found out, or imagined, about cycles, celestial or terrestial, these linear concepts were the elements that dominated the thought of Confucian scholars and Taoist peasant-farmers alike.

Strange as it may seem to those who still think in terms of the "timeless Orient," the culture of China was, on the whole, more of the Iranic, Judaeo-Christian than of the Indo-Hellenic types. The conclusion then springs to the mind: If Chinese civilisation did not spontaneously develop modern natural science as Western Europe did, though China had been much more advanced in the fifteen pre-Renaissance centuries, it had nothing to do with China's attitude toward time. Other ideological factors remain for scrutiny, apart from the concrete geographical, social, and economic conditions and structures which may yet surface to bear the main burden of the explanation.

[a] 商

Table of Chinese Dynasties

夏 Hsia kingdom (legendary?)	*c.* −2000 to *c.* −1520
商 Shang (Yin) kingdom	*c.* −1520 to *c.* −1030

周 Chou dynasty (Feudal Age)	⎰ Early Chou period	*c.* −1030 to −722
	Ch'un Ch'iu period 春秋	−722 to −480
	Warring States (Chan Kuo) period 戰國	−480 to −221

First Unification

秦 Ch'in dynasty	−221 to −207

漢 Han dynasty	Ch'ien Han (Earlier or Western)	−202 to +9
	Hsin interregnum	+9 to +23
	Hou Han (Later or Eastern)	+25 to +220

First Partition

三國 San Kuo (Three Kingdoms period)	+221 to +265
蜀 Shu (Han)	+221 to +264
魏 Wei	+220 to +265
吳 Wu	+222 to +280

Second Unification

晉 Chin dynasty: Western	+265 to +317
Eastern	+317 to +420
劉宋 (Liu) Sung dynasty	+420 to +479

Second Partition

Northern and Southern Dynasties (Nan Pei ch'ao)

齊 Ch'i dynasty	+479 to +502
梁 Liang dynasty	+502 to +557
陳 Ch'ên dynasty	+557 to +589
魏 ⎰ Northern (T'opa) Wei dynasty	+386 to +535
Western (T'opa) Wei dynasty	+535 to +556
Eastern (T'opa) Wei dynasty	+534 to +550
北齊 Northern Ch'i dynasty	+550 to +577
北周 Northern Chou (Hsienpi) dynasty	+557 to +581

Third Unification

隋 Sui dynasty	+581 to +618
唐 T'ang dynasty	+618 to +906

Third Partition

五代 Wu Tai (Five Dynasties period) [Later   + 907 to   + 960
Liang, Later T'ang (Turkic), Later Chin
(Turkic), Later Han (Turkic) and Later
Chou]
遼   Liao (Ch'itan Tartar) dynasty   + 907 to + 1124
西遼 West Liao dynasty (Qasrā-Khiṭāi)   + 1124 to + 1211
西夏 Hsi Hsia (Tangut Tibetan) state   + 986 to + 1227

Fourth Unification

宋   Northern Sung dynasty   + 960 to + 1126
宋   Southern Sung dynasty   + 1127 to + 1279
金   Chin (Jurchen Tartar) dynasty   + 1115 to + 1234
元   Yüan (Mongol) dynasty   + 1260 to + 1368
明   Ming dynasty   + 1368 to + 1644
清   Ch'ing (Manchu) dynasty   + 1644 to + 1911
民國 Republic   + 1912

N.B.   When no modifying term in brackets is given, the dynasty was purely
Chinese. During the Eastern Chin period there were no less than eighteen independent States (Hunnish, Tibetan, Hsienpi, Turkic, etc.) in the north. The
term "Liu ch'ao" (Six Dynasties) is often used by historians of literature. It refers
to the south and covers the period from the beginning of the + 3rd to the end of
the + 6th centuries, including (San Kuo) Wu, Chin, (Liu) Sung, Ch'i, Liang, and
Ch'ên.